RAY'S™

NEW PRIMARY

ARITHMETIC

FOR

YOUNG LEARNERS

Originally published by
Van Antwerp, Bragg & Co.

This edition published by

PREFACE

The remarkable and long continued popularity of *Ray's Arithmetics* has induced the publishers to present them to the public in a revised form, as *Ray's New Arithmetics*.

(1) To present the books in improved type, with better arrangement, and in a more pleasing outward dress.

(2) To introduce such new features as will adapt the series more perfectly to the present methods of instruction.

The friends of the series can best judge what success in attaining these objects has been made.

The publishers hereby express their gratitude to many prominent educators who have contributed to this revision, and only regret that their number prevents the mention of names.

Cincinnati, April, 1877.

1130 Fenway Circle, Fenton, Michigan 48430
www.mottmedia.com

ISBN-10: 0-88062-059-5
ISBN-13: 978-0-88062-059-8
Printed by Color House Graphics, Grand Rapids, Michigan, USA.

PRESENT PUBLISHER'S PREFACE

We are honored and happy to bring to you the classic Arithmetics by Joseph Ray. In the 1800s these popular books sold more than any other arithmetics in America, in fact over 120,000,000 copies. Now with this reprinting, they are once again available for America's students.

Ray's Arithmetics are organized in an orderly manner around the discipline of arithmetic itself. They present principles and follow up each one with examples which include difficult problems to challenge the best students. Students who do not master a concept the first time can return to it later, work the more difficult problems, and master the concept. Thus in these compact volumes is a complete arithmetic course to study in school, to help in preparing for ACT and SAT tests, and to use for reference throughout a lifetime.

In order to capture the spirit of the original Ray's, we have refrained from revising the problems and prices. Only a few words have been changed, as we felt it wise. Thus students will have to rely on their arithmetic ability to solve the problems. Also the charm of a former era lives on in this reprinting. Flour and salt are sold by the barrel, kegs may contain tar, and postage stamps cost 3¢ each. Through this content, students learn social history of the 1800s in a unique, hands-on manner at the same time they are mastering arithmetic.

The series consists of four books ranging from Primary Arithmetic to Higher Arithmetic, as well as answer keys to accompany them. We have added a teacher's guide to help today's busy teachers and parents.

We wish to express our appreciation to the staff of the Special Collections Library at Miami University, Oxford, Ohio, for its cooperation in allowing us to use copies of their original Ray's Arithmetics.

George M. Mott, Founder
Mott Media

SUGGESTIONS TO TEACHERS

In beginning the study of Arithmetic, the first step for pupils to learn is to count readily. This is not mastered without much practice in counting *objects*. Movable objects are better for exercises in counting than pictures. Some objects of this kind should always be kept in the school-room,—such as marbles, beans, kernels of corn, or pebbles.

The second step is to combine numbers. To master the different combinations to 20, the pupils should first be taught to write the tables corresponding with those in the book, either upon their slates or on the blackboard, during the recitation. This will prevent counting upon the fingers, a habit difficult to overcome when once acquired.

As the abstract exercises in this book, up to 20, are exhaustive in Addition and Subtraction, and as complete in Multiplication and Division as possible in order to secure variety, it would be well to prepare additional concrete examples from day to day to correspond with the very full abstract exercises. An excellent practice is to require each pupil to bring two or more concrete examples of his own to each recitation.

Teach one thing at a time, and teach it thoroughly.

LESSON I.

NOTE.—This lesson is intended to suggest how to teach the child to count objects and to express their number by figures. It comprises the first ten numbers. The pupil points to each ball and says, *one; one, two; one, two, three,* etc. Then the teacher directs him to write the figure for *one,* for *two,* etc.

. . . . one, . 1, *1.*
. . . . two, . 2, *2.*
. . . . three, . 3, *3.*
. . . . four, . 4, *4.*
. . . . five, . 5, *5.*
. . . . six, . 6, *6.*
. . . . . . seven, . 7, *7.*
. . . . . eight, . 8, *8.*
. . . nine, . 9, *9.*
. ten, . 10, *10.*

LESSON II.

NOTE.—In this lesson the numbering of objects is extended to 40. A single column or less may constitute one exercise, as may seem best to the teacher.

Eleven	11	Twenty-one	21	Thirty-one	31
Twelve	12	Twenty-two	22	Thirty-two	32
Thirteen	13	Twenty-three	23	Thirty-three	33
Fourteen	14	Twenty-four	24	Thirty-four	34
Fifteen	15	Twenty-five	25	Thirty-five	35
Sixteen	16	Twenty-six	26	Thirty-six	36
Seventeen	17	Twenty-seven	27	Thirty-seven	37
Eighteen	18	Twenty-eight	28	Thirty-eight	38
Nineteen	19	Twenty-nine	29	Thirty-nine	39
TWENTY	20	THIRTY	30	FORTY	40

LESSON III.

NOTE.—In this lesson the numbering of objects is extended to 70. Each exercise should include a review of the preceding ones.

Forty-one	41	Fifty-one	51	Sixty-one	61
Forty-two	42	Fifty-two	52	Sixty-two	62
Forty-three	43	Fifty-three	53	Sixty-three	63
Forty-four	44	Fifty-four	54	Sixty-four	64
Forty-five	45	Fifty-five	55	Sixty-five	65
Forty-six	46	Fifty-six	56	Sixty-six	66
Forty-seven	47	Fifty-seven	57	Sixty-seven	67
Forty-eight	48	Fifty-eight	58	Sixty-eight	68
Forty-nine	49	Fifty-nine	59	Sixty-nine	69
FIFTY	50	SIXTY	60	SEVENTY	70

LESSON IV.

Seventy-one	71	Eighty-one	81	Ninety-one	91
Seventy-two	72	Eighty-two	82	Ninety-two	92
Seventy-three	73	Eighty-three	83	Ninety-three	93
Seventy-four	74	Eighty-four	84	Ninety-four	94
Seventy-five	75	Eighty-five	85	Ninety-five	95
Seventy-six	76	Eighty-six	86	Ninety-six	96
Seventy-seven	77	Eighty-seven	87	Ninety-seven	97
Seventy-eight	78	Eighty-eight	88	Ninety-eight	98
Seventy-nine	79	Eighty-nine	89	Ninety-nine	99
EIGHTY	80	NINETY	90	ONE HUNDRED	100

LESSON V.

NOTE.—Pupils should be taught to *read* figures readily from 1 to 100. The figures should be copied on the blackboard.

NUMBERS TO BE READ.

0	23	50	61	71	80	78	59
1	32	15	26	27	18	87	95
10	33	51	62	72	81	88	69
11	4	25	36	37	28	9	96
2	40	52	63	73	82	90	79
20	14	35	46	47	38	19	97
12	41	53	64	74	83	91	89
21	24	45	56	57	48	29	98
22	42	54	65	75	84	92	99
3	34	55	66	67	58	39	100
30	43	6	7	76	85	93	
13	44	60	70	77	68	49	
31	5	16	17	8	86	94	

LESSON VI.

NOTE.—The pupils must be thoroughly exercised in *writing* numbers. One or more pupils at a time may be sent to the blackboard, or the work may be done at their seats with pencil and slate.

NUMBERS TO BE WRITTEN.

1. Naught; one, ten; two, twenty; three, thirty; four, forty; five, fifty; six, sixty; seven, seventy; eight, eighty; nine, ninety.

2. Eleven; twelve, twenty-one; thirteen, thirty-one; fourteen, forty-one; fifteen, fifty-one; sixteen, sixty-one; seventeen, seventy-one; eighteen, eighty-one; nineteen, ninety-one.

3. Twenty-two; twenty-three, thirty-two; twenty-four, forty-two; twenty-five, fifty-two; twenty-six, sixty-two; twenty-seven, seventy-two; twenty-eight, eighty-two; twenty-nine, ninety-two.

4. Thirty-three; thirty-four, forty-three; thirty-five, fifty-three; thirty-six, sixty-three; thirty-seven, seventy-three; thirty-eight, eighty-three; thirty-nine, ninety-three.

5. Forty-four; forty-five, fifty-four; forty-six, sixty-four; forty-seven, seventy-four; forty-eight, eighty-four; forty-nine, ninety-four.

6. Fifty-five; fifty-six, sixty-five; fifty-seven, seventy-five; fifty-eight, eighty-five; fifty-nine, ninety-five.

7. Sixty-six; sixty-seven, seventy-six; sixty-eight, eighty-six; sixty-nine, ninety-six.

8. Seventy-seven; seventy-eight, eighty-seven; seventy-nine, ninety-seven.

9. Eighty-eight; eighty-nine, ninety-eight.

10. Ninety-nine. One hundred.

LESSON VII.

NOTE.—These exercises are intended for use with the Numeral Frame or with counters of some kind,—marbles, pebbles, kernels of corn, beans, or bits of pasteboard. The objects should be arranged in distinct groups, to represent each number indicated.

1. How many counters have we here? (/)

2. How many are 1 and 1? One taken away from 2 leaves how many? How many ones in 2? How many are two times 1?

3. How many are 2 and 1? How many are 1 and 1 and 1? How many are three times 1?

4. One taken away from 3 leaves how many? Two taken away from 3 leaves how many? How many ones in 3?

5. How many are 3 and 1? How many are 2 and 2? How many are 1 and 1 and 1 and 1? How many are four times 1? How many are two times 2?

6. One taken from 4 leaves how many? Two from 4

leaves how many? Three from 4 leaves how many? How many ones in 4? How many twos in 4?

7. How many are 4 and 1? How many are 3 and 2? How many are 1 and 1 and 1 and 1 and 1? How many are five times 1?

8. One from 5 leaves how many? Two from 5 leaves how many? Three from 5 leaves how many? Four from 5 leaves how many? How many ones in 5?

LESSON VIII.

1. How many are 5 and 1? How many are 4 and 2? How many are 3 and 3? How many are six times 1? How many are three times 2? How many are two times 3?

2. One from 6 leaves how many? Two from 6? Three from 6? Four from 6? Five from 6? How many ones in 6? How many twos in 6? How many threes in 6?

3. How many are 6 and 1? How many are 5 and 2? How many are 4 and 3? How many are 3 and 4? How many are seven times 1?

4. One from 7 leaves how many? Two from 7? Three from 7? Four from 7? Five from 7? Six from 7? How many ones in 7?

5. How many are 7 and 1? How many are 6 and 2? How many are 5 and 3? How many are 4 and 4? How many are 3 and 5? How many are 2 and 6?

6. How many are eight times 1? How many are four times 2? How many are two times 4?

7. One from 8 leaves how many? Two from 8? Three from 8? Four from 8? Five from 8? Six from 8? Seven from 8?

8. How many ones in 8? How many twos in 8? How many fours in 8?

LESSON IX.

1. How many are 8 and 1? How many are 7 and 2? How many are 6 and 3? How many are 5 and 4? How many are 4 and 5? How many are 3 and 6? How many are 2 and 7?

2. How many are nine times 1? How many are three times 3?

3. One from 9 leaves how many? Two from 9? Three from 9? Four from 9? Five from 9? Six from 9? Seven from 9? Eight from 9?

4. How many ones in 9? How many threes in 9?

5. How many are 9 and 1? How many are 8 and 2? How many are 7 and 3? How many are 6 and 4? How many are 5 and 5?

6. How many are 2 and 8? How many are 3 and 7? How many are 4 and 6?

7. How many are ten times 1? How many are five times 2? How many are two times 5?

8. One from 10 leaves how many? Two from 10? Three from 10? Four from 10? Five from 10? Six from 10? Seven from 10? Eight from 10? Nine from 10?

9. How many ones in 10? How many twos in 10? How many fives in 10?

LESSON X.

In the picture how many birds are sitting on the bush? How many in the flock that seems to be lighting? There are two distant flocks flying: how many birds in each flock?

1. How many birds are two birds and five birds? How many birds are seven birds and four birds?

2. How many are 2 and 5? 7 and 4?

3. How many birds are two birds and four birds? How many are five birds and seven birds?

4. How many are 2 and 4? 5 and 7?

5. There are three flowers on one branch and three on another: how many flowers on both branches?

LESSON XI.

1 and 1 are 2	6 and 1 are 7
2 and 1 are 3	1 and 6 are 7
1 and 2 are 3	7 and 1 are 8
3 and 1 are 4	1 and 7 are 8
1 and 3 are 4	8 and 1 are 9
4 and 1 are 5	1 and 8 are 9
1 and 4 are 5	9 and 1 are 10
5 and 1 are 6	1 and 9 are 10
1 and 5 are 6	10 and 1 are 11

1 and 10 are 11.

1. Francis had 2 cents, and his mother gave him 1 cent more: how many had he then?

SOLUTION.—Francis had then 2 cents and 1 cent, which are 3 cents.

2. John had 1 raisin, and his sister gave him 3 raisins more: how many had he then?

3. Mary had 4 pears, and her mother gave her 1 pear more: how many had she then?

4. Jane had 1 cherry, and her brother gave her 5 cherries more: how many did she then have?

5. George has 6 cents, and John has 1 cent: how many cents have both?

6. William had 1 plum, and his cousin gave him 7 plums more: how many had he then?

7. There were 8 oranges on a dish, and 1 more orange was placed on it: how many were then on the dish?

8. Henry had 1 peach, and his mother gave him 9 more: how many had he then?

9. How many are 10 cents and 1 cent?

LESSON XII.

2 and 1 are 3	2 and 6 are 8
1 and 2 are 3	6 and 2 are 8
2 and 2 are 4	2 and 7 are 9
2 and 3 are 5	7 and 2 are 9
3 and 2 are 5	2 and 8 are 10
2 and 4 are 6	8 and 2 are 10
4 and 2 are 6	2 and 9 are 11
2 and 5 are 7	9 and 2 are 11
5 and 2 are 7	2 and 10 are 12

10 and 2 are 12.

1. Mary had two birds, and a friend gave her 2 more: how many birds had she then?

Solution.—Mary had then 2 birds and 2 birds, which are 4 birds.

2. Daniel has 3 tops, and Francis has 2: how many tops have they both?
3. John had 2 chestnuts, and found 4 more: how many did he then have?
4. Helen had 5 apples, and her brother gave her 2 more: how many had she then?
5. Ellen had 2 chickens, and her cousin gave her 6 more: how many had she then?
6. John had 2 cakes, and his mother gave him 7 more: how many did he then have?
7. Frank had 8 marbles, and found 2 more: how many did he then have?
8. Harry caught 2 fishes, and Edward caught 9: how many did both catch?
9. How many are 10 cents and 2 cents?

LESSON XIII.

3 and 1 are 4	3 and 6 are 9
1 and 3 are 4	6 and 3 are 9
3 and 2 are 5	3 and 7 are 10
2 and 3 are 5	7 and 3 are 10
3 and 3 are 6	3 and 8 are 11
3 and 4 are 7	8 and 3 are 11
4 and 3 are 7	3 and 9 are 12
3 and 5 are 8	9 and 3 are 12
5 and 3 are 8	3 and 10 are 13

10 and 3 are 13.

1. Julius had 3 cents, and he found 2 more: how many cents had he then?

SOLUTION.—Julius had then 3 cents and 2 cents, which are 5 cents.

2. Francis has 3 dimes in his hand, and 3 in his pocket: how many dimes has he?

3. Emma has 4 apples: if her mother give her 3 more, how many apples will she have?

4. There are 3 pears on one limb, and 5 on another: how many pears on both limbs?

5. Mary has 6 pens, and Belle has 3: how many pens have both?

6. Henry has 3 books, and Oliver has 7: how many books have both?

7. Charles caught 8 rabbits, and Samuel caught 3: how many did both catch?

8. How many are 3 cents and 9 cents?

9. My pencil cost 10 cents, and my pen 3 cents: how much did both cost?

LESSON XIV.

4 and 1 are 5	4 and 6 are 10
1 and 4 are 5	6 and 4 are 10
4 and 2 are 6	4 and 7 are 11
2 and 4 are 6	7 and 4 are 11
4 and 3 are 7	4 and 8 are 12
3 and 4 are 7	8 and 4 are 12
4 and 4 are 8	4 and 9 are 13
4 and 5 are 9	9 and 4 are 13
5 and 4 are 9	4 and 10 are 14

10 and 4 are 14.

1. James had 4 pens, and he found 2 more: how many had he then?

SOLUTION.—James had then 4 pens and 2 pens, which are 6 pens.

2. Mary has 3 pins in one hand, and 4 in the other: how many pins has she in both?
3. Francis has 4 chestnuts in his hand, and 4 in his pocket: how many has he in all?
4. There are 5 horses in 1 field, and 4 in another: how many are there in both fields?
5. Cora spent 4 cents for tape, and 6 cents for ribbon: how many cents did she spend?
6. I had 7 apples, and bought 4 more: how many did I then have?
7. If a lemon cost 4 cents, and an orange 8 cents, how much will both cost?
8. I sold a calf for 9 dollars, and a sheep for 4 dollars: how much did I get for both?
9. How many are 4 cents and 10 cents?

LESSON XV.

5 and 1 are 6	5 and 6 are 11
1 and 5 are 6	6 and 5 are 11
5 and 2 are 7	5 and 7 are 12
2 and 5 are 7	7 and 5 are 12
5 and 3 are 8	5 and 8 are 13
3 and 5 are 8	8 and 5 are 13
5 and 4 are 9	5 and 9 are 14
4 and 5 are 9	9 and 5 are 14
5 and 5 are 10	5 and 10 are 15

10 and 5 are 15.

1. A hen has 5 black chickens and 2 white ones: how many chickens has she?

SOLUTION.—She has 5 chickens and 2 chickens, which are 7 chickens.

2. How many are 3 square blocks and 5 square blocks?
3. I gave 5 cents for a whistle, and 4 cents for a top: how much did I give for both?
4. Emma had 5 cakes, and her mother gave her 5 more: how many had she then?
5. There are 6 chairs in one room, and 5 in another: how many chairs in both rooms?
6. There are 5 boys in one class, and 7 in another: how many are there in both classes?
7. If a lemon cost 8 cents, and an orange 5 cents, how much will both cost?
8. There are 5 letters in my name, and 9 in yours: how many letters in both names?
9. If you put 10 balls by the side of 5 balls, how many balls will there be?

LESSON XVI.

6 and 1 are 7	5 and 6 are 11
1 and 6 are 7	6 and 6 are 12
6 and 2 are 8	6 and 7 are 13
2 and 6 are 8	7 and 6 are 13
6 and 3 are 9	6 and 8 are 14
3 and 6 are 9	8 and 6 are 14
6 and 4 are 10	6 and 9 are 15
4 and 6 are 10	9 and 6 are 15
6 and 5 are 11	6 and 10 are 16

10 and 6 are 16.

1. A farmer has 6 cows in one field, and 2 in another: how many cows in both fields?

SOLUTION.—The farmer has 6 cows and 2 cows, which are 8 cows.

2. James has 3 marbles in one pocket, and 6 in another how many has he in both?
3. Francis had 6 cents, and Mary, 4 cents: how many cents had both?
4. There are 5 pigs in one pen, and 6 in another: how many pigs in both pens?
5. If you have 6 plums in each hand, how many plums will you have in both hands?
6. Lucy gave 7 cents to one poor man, and 6 cents to another: how many cents did she give to both?
7. A man had 6 horses, and bought 8 more: how many horses did he then have?
8. A lady traveled 9 miles by water, and 6 miles by land: how far did she travel?
9. How many are 10 days and 6 days:

LESSON XVII.

7 and 1 are 8	5 and 7 are 12
1 and 7 are 8	7 and 6 are 13
7 and 2 are 9	6 and 7 are 13
2 and 7 are 9	7 and 7 are 14
7 and 3 are 10	7 and 8 are 15
3 and 7 are 10	8 and 7 are 15
7 and 4 are 11	7 and 9 are 16
4 and 7 are 11	9 and 7 are 16
7 and 5 are 12	7 and 10 are 17

10 and 7 are 17.

1. If you place 7 marbles by the side of 2 marbles, how many will there be altogether?

SOLUTION.—There will be 7 marbles and 2 marbles, which are 9 marbles.

2. There are 3 sheep in one field, and 7 in another: how many sheep in both fields?
3. There are 7 boys on one bench, and 4 on another: how many are there on both benches?
4. There are 5 chairs in one room, and 7 in another: how many chairs in both rooms?
5. Thomas had 7 apples, and his mother gave him 6 more: how many had he then?
6. I bought a melon for 7 cents, and a squash for 7 cents: how much did both cost?
7. I paid 8 cents for a slate, and 7 cents for some pencils: how many cents did I spend?
8. Fanny had 7 roses, and she plucked 9 more: how many had she then?
9. How many are 10 dollars and 7 dollars?

LESSON XVIII.

8 and 1 are 9	5 and 8 are 13
1 and 8 are 9	8 and 6 are 14
8 and 2 are 10	6 and 8 are 14
2 and 8 are 10	8 and 7 are 15
8 and 3 are 11	7 and 8 are 15
3 and 8 are 11	8 and 8 are 16
8 and 4 are 12	8 and 9 are 17
4 and 8 are 12	9 and 8 are 17
8 and 5 are 13	8 and 10 are 18

10 and 8 are 18.

1. James has 8 nuts in his pocket, and 2 in his hand: how many nuts has he?

SOLUTION.—James has 8 nuts and 2 nuts, which are 10 nuts.

2. Mary has 3 pins in one hand, and 8 in the other: how many has she in both hands?
3. There are 8 geese in one pond, and 4 in another: how many geese in both ponds?
4. Thomas had 5 marbles, and has found 8 more: how many marbles has he now?
5. Harvey found 8 eggs, and Thomas 6: how many eggs did both find?
6. I gave 7 dollars for a vest, and 8 dollars for a coat: how much did both cost?
7. I bought 8 yards of blue cloth, and 8 yards of black: how many yards did I buy altogether?
8. Anna is 9 years old, and Alice is 8 years older than Anna: how old is Alice?
9. How many are 8 cents and 10 cents?

LESSON XIX.

9 and 1 are 10	5 and 9 are 14
1 and 9 are 10	9 and 6 are 15
9 and 2 are 11	6 and 9 are 15
2 and 9 are 11	9 and 7 are 16
9 and 3 are 12	7 and 9 are 16
3 and 9 are 12	9 and 8 are 17
9 and 4 are 13	8 and 9 are 17
4 and 9 are 13	9 and 9 are 18
9 and 5 are 14	9 and 10 are 19

10 and 9 are 19.

1. A cat caught 9 mice one day, and 2 the next: how many did she catch in both days?

SOLUTION.—She caught 9 mice and 2 mice, which are 11 mice.

2. Mary gave 3 cents for paper, and 9 cents for a book: how much did both cost?
3. Joseph caught 9 fishes in one pool, and 4 in another: how many did he catch?
4. If five horses are in one field, and 9 in another, how many horses are in both fields?
5. If you have 9 oranges, and buy 6 more, how many oranges will you then have?
6. Charles had 7 plums, and John gave him 9: how many did he then have?
7. Sarah had 9 buttons, and her aunt gave her 8 more: how many had she then?
8. George and Henry have 9 cents each: how many cents have both?
9. How many are 10 pounds and 9 pounds?

LESSON XX.

10 and 1 are 11	5 and 10 are 15
1 and 10 are 11	10 and 6 are 16
10 and 2 are 12	6 and 10 are 16
2 and 10 are 12	10 and 7 are 17
10 and 3 are 13	7 and 10 are 17
3 and 10 are 13	10 and 8 are 18
10 and 4 are 14	8 and 10 are 18
4 and 10 are 14	10 and 9 are 19
10 and 5 are 15	9 and 10 are 19

10 and 10 are 20.

1. I paid 10 cents for ink, and 2 cents for paper: how much did I pay for both?

SOLUTION.—I paid for both 10 cents and 2 cents, which are 12 cents.

2. Mary gave 3 dollars for a dress, and 10 dollars for a shawl: what did she give for both?

3. There are 10 trees in one row, and 4 in another: how many trees in both rows?

4. In one pasture there are 5 cows, and in another 10: how many cows in both pastures?

5. I bought 10 yards of blue ribbon, and 6 yards of white: how many yards of ribbon did I buy?

6. There are 7 girls in one class, and 10 in another: how many girls in both classes?

7. I received 10 dollars for peaches, and 8 dollars for plums: how much in all?

8. Edwin found 9 nuts under one tree, and 10 under another: how many nuts did he find?

9. Mary had 10 cents, and her mother gave her 10 more: how many did she then have?

LESSON XXI.

REVIEW.

1. How many are 2 and 4?

SOLUTION.—2 and 4 are 6.

2. How many are 3 and 5? 4 and 7?
3. How many are 6 and 4? 3 and 8?
4. How many are 2 and 9? 8 and 2?
5. How many are 2 and 10? 7 and 2?
6. How many are 3 and 3? 4 and 4?
7. How many are 5 and 4? 6 and 5?
8. How many are 2 and 5? 3 and 10?
9. How many are 10 and 4? 3 and 9?
10. How many are 6 and 7? 7 and 5?
11. How many are 4 and 8? 5 and 9?

12. How many are 10 and 6? 9 and 7?
13. How many are 7 and 7? 6 and 6?
14. How many are 8 and 6? 10 and 7?
15. How many are 2 and 6? 6 and 9?
16. How many are 9 and 4? 10 and 9?
17. How many are 10 and 8? 5 and 8?
18. How many are 8 and 7? 2 and 2?
19. How many are 5 and 5? 7 and 3?
20. How many are 8 and 8? 9 and 9?
21. How many are 4 and 3? 2 and 3?
22. How many are 5 and 10? 8 and 9?

23. How many are 2 and 2 and 2?

SOLUTION.—2 and 2 are 4; 4 and 2 are 6.

24. How many are 4 and 4 and 4? 5 and 5 and 5?
25. How many are 6 and 4 and 6? 2 and 6 and 8?

26. How many are 3 and 5 and 7? 4 and 5 and 9?
27. How many are 6 and 2 and 9? 5 and 2 and 3?
28. How many are 3 and 5 and 9? 4 and 4 and 6?
29. How many are 4 and 6 and 7? 3 and 8 and 9?
30. How many are 2 and 3 and 2? 2 and 6 and 7?

LESSON XXII.

REVIEW.

1. How many are 2 and 4 and 2? 2 and 7 and 7?
2. How many are 3 and 5 and 5? 4 and 5 and 8?
3. How many are 2 and 5 and 2? 3 and 6 and 5?
4. How many are 6 and 2 and 2? 2 and 2 and 7?
5. How many are 4 and 4 and 2? 3 and 3 and 5?
6. How many are 3 and 6 and 3? 2 and 4 and 6?
7. How many are 3 and 4 and 5? 5 and 2 and 7?
8. How many are 2 and 4 and 7? 3 and 8 and 8?
9. How many are 5 and 6 and 8? 4 and 6 and 9?
10. How many are 3 and 6 and 6? 2 and 7 and 9?

11. How many are 2 and 3 and 6? 2 and 7 and 3?
12. How many are 2 and 9 and 9? 3 and 6 and 9?
13. How many are 5 and 3 and 8? 3 and 3 and 8?
14. How many are 2 and 2 and 9? 2 and 8 and 2?
15. How many are 8 and 2 and 8? 5 and 2 and 5?
16. How many are 4 and 2 and 9? 6 and 6 and 7?
17. How many are 5 and 4 and 4? 5 and 5 and 9?
18. How many are 7 and 4 and 9? 2 and 7 and 8?
19. How many are 4 and 3 and 2? 2 and 3 and 3?
20. How many are 6 and 7 and 7? 4 and 8 and 8?

21. How many are 2 and 3 and 8? 2 and 9 and 8?
22. How many are 6 and 2 and 6? 3 and 6 and 7?

23. How many are 4 and 5 and 5? 4 and 5 and 6?
24. How many are 4 and 6 and 8? 3 and 7 and 9?
25. How many are 3 and 3 and 4? 5 and 8 and 7?
26. How many are 2 and 3 and 9? 3 and 3 and 7?
27. How many are 5 and 5 and 6? 5 and 5 and 8?
28. How many are 4 and 7 and 8? 3 and 6 and 8?
29. How many are 3 and 3 and 3? 4 and 2 and 5?
30. How many are 4 and 7 and 7? 4 and 4 and 9?

31. How many are 4 and 5 and 7?

SOLUTION.—4 and 5 are 9, and 7 are 16.

32. How many are 3 and 7 and 7? 2 and 4 and 8?
33. How many are 5 and 6 and 7? 3 and 4 and 4?
34. How many are 3 and 7 and 8? 2 and 5 and 9?
35. How many are 5 and 6 and 9? 2 and 5 and 8?
36. How many are 3 and 3 and 9? 6 and 4 and 3?
37. How many are 3 and 4 and 7? 4 and 4 and 7?
38. How many are 5 and 7 and 7? 3 and 4 and 8?
39. How many are 6 and 6 and 8? 4 and 4 and 8?
40. How many are 2 and 5 and 6? 3 and 4 and 9?

41. Begin with 1 and count by 2's to 19.

SOLUTION.—1, 3, 5, 7, 9, 11, 13, 15, 17, 19.

42. Begin with 1 and count by 3's to 19.
43. Begin with 1 and count by 4's to 17.
44. Begin with 1 and count by 5's to 16.
45. Count by 2's to 20. Count by 3's to 18.
46. Count by 4's to 20. Count by 5's to 20.
47. Begin with 2 and count by 3's to 20.
48. Begin with 2 and count by 4's to 18.
49. Begin with 2 and count by 5's to 17.
50. Begin with 3 and count by 4's to 19.

LESSON XXIII.

1. Mary paid 5 cents for ribbon, 4 cents for thread, and 3 cents for tape: how much did she pay for all?

SOLUTION.—Mary paid for all 5 cents and 4 cents and 3 cents, which are 12 cents.

2. Joseph caught 6 fishes, Samuel 3, and Henry 5: how many fishes did they all catch?

3. A boy spent 7 cents for candy, 3 cents for cakes, and 6 cents for apples: how much did he spend?

4. Six peaches and 3 peaches and 6 peaches are how many peaches?

5. Eight dollars and two dollars and eight dollars are how many dollars?

6. Jane has 2 pins, and Mary gives her 4; she then finds 7 more: how many pins has she?

7. James has 4 marbles: he buys 6 at a store, and Edward gives him 7 more: how many marbles has he then?

8. I buy a hat for 4 dollars, a vest for 5 dollars, and a coat for 10 dollars, when I find that I have spent all my money: how much money had I?

9. A farmer sold a barrel of apples for three dollars, a tub of butter for nine dollars, and a load of wood for seven dollars: how much did he receive for all?

10. I buy four apples for 7 cents, two pears for 3 cents, and five oranges for 10 cents: how much do I spend?

11. I buy 2 apples of one man for 2 cents, 6 of another for 5 cents, and 7 of another for 10 cents: how many apples do I buy? How many cents do I spend?

12. John bought 5 pears for 6 cents, 3 pears for 4 cents, and 2 pears for 5 cents: how many pears did he buy? How much did they all cost?

LESSON XXIV.

Here is a picture of some owls sitting on the branch of a pine tree. How many owls are there? There are also some bats flying: how many bats can you see?

1. There were three owls sitting on a tree; two of them flew away, how many were left?

2. Three less 2 are how many?

3. Seven bats were seen flying, four of them flew into a barn: how many were still outside?

4. Seven less 4 are how many?

5. There were nine leaves on one twig, but three of them fell off: how many leaves were left?

6. Nine less 3 are how many?

LESSON XXV.

1 from 1 leaves 0	1 from 6 leaves 5
1 from 2 leaves 1	6 from 7 leaves 1
2 from 3 leaves 1	1 from 7 leaves 6
1 from 3 leaves 2	7 from 8 leaves 1
3 from 4 leaves 1	1 from 8 leaves 7
1 from 4 leaves 3	8 from 9 leaves 1
4 from 5 leaves 1	1 from 9 leaves 8
1 from 5 leaves 4	9 from 10 leaves 1
5 from 6 leaves 1	1 from 10 leaves 9

1. Mary had 3 roses, and she gave 2 of them to Henry: how many had she left?

SOLUTION.—Mary had left 3 roses less 2 roses, which is 1 rose.

2. If 1 melon be taken from 4 melons, how many melons will remain?

3. If you have 5 nuts, and give away 4 of them, how many will you then have?

4. Thomas had 6 pigeons, but one of them died: how many were left?

5. Charles had 7 marbles, and lost 6 of them: how many marbles had he left?

6. Alice had 8 chickens, and 1 was killed: how many had she then?

7. Jane had 9 cents, and spent 8 cents: how many cents had she left?

8. If you take 9 apples from 10 apples, how many will remain?

9. Ella had 10 plums, and she gave 1 to her sister: how many plums had she left?

LESSON XXVI.

2 from 2 leaves 0	5 from 7 leaves 2
2 from 3 leaves 1	2 from 8 leaves 6
1 from 3 leaves 2	6 from 8 leaves 2
2 from 4 leaves 2	2 from 9 leaves 7
2 from 5 leaves 3	7 from 9 leaves 2
3 from 5 leaves 2	2 from 10 leaves 8
2 from 6 leaves 4	8 from 10 leaves 2
4 from 6 leaves 2	2 from 11 leaves 9
2 from 7 leaves 5	9 from 11 leaves 2

1. Frank had 3 apples, and gave 1 to his brother: how many had he left?

SOLUTION.—Frank had left 3 apples less 1 apple, which are 2 apples.

2. William had 4 cents: after spending 2 cents for nuts, how many had he left?

3. There were 5 birds in a tree: after 2 of them flew away, how many were left?

4. Daniel caught 6 mice, but 2 of them got away: how many were left?

5. Francis bought 7 oranges, and ate 2 of them: how many had he then?

6. John had 8 marbles, but he lost 2 of them: how many remained.

7. A hen had 9 chickens, but 2 of them died: how many were left?

8. Margaret had 10 cakes, and gave 8 of them to her sister: how many had she left?

9. If 2 yards be taken from 11 yards, how many yards will remain?

LESSON XXVII.

3 from 3 leaves 0	5 from 8 leaves 3
3 from 4 leaves 1	3 from 9 leaves 6
1 from 4 leaves 3	6 from 9 leaves 3
3 from 5 leaves 2	3 from 10 leaves 7
2 from 5 leaves 3	7 from 10 leaves 3
3 from 6 leaves 3	3 from 11 leaves 8
3 from 7 leaves 4	8 from 11 leaves 3
4 from 7 leaves 3	3 from 12 leaves 9
3 from 8 leaves 5	9 from 12 leaves 3

1. Mary had 4 oranges: after eating 3, how many did she have left?

Solution.—Mary had left 4 oranges less 3 oranges, which is 1 orange.

2. Lucy had 5 ducks, but 2 of them died: how many had she then?
3. Six persons are in a carriage: if 3 of them get out, how many will remain?
4. Cora had 7 pins, and lost 4 of them: how many pins did she then have?
5. Francis had 8 cents, and spent 3 of them: how many cents had he left?
6. A hen had 9 chickens, but a hawk carried off 6 of them: how many were left?
7. John had 10 lemons, and gave 3 to a poor sick boy: how many had he left?
8. If 8 be taken from 11, how many will remain?
9. Eliza had 12 cents, and spent 3 cents: how many cents had she left?

LESSON XXVIII.

4 from 4 leaves 0	4 from 9 leaves 5
1 from 5 leaves 4	6 from 10 leaves 4
4 from 5 leaves 1	4 from 10 leaves 6
2 from 6 leaves 4	7 from 11 leaves 4
4 from 6 leaves 2	4 from 11 leaves 7
3 from 7 leaves 4	8 from 12 leaves 4
4 from 7 leaves 3	4 from 12 leaves 8
4 from 8 leaves 4	9 from 13 leaves 4
5 from 9 leaves 4	4 from 13 leaves 9

1. Henry has 5 apples: if he eat 4 of them, how many will he have left?

SOLUTION.—Henry will have left 5 apples less 4 apples, which is 1 apple.

2. Eliza had 6 birds in a cage: she let 2 of them out: how many remained in the cage?

3. There were 7 ducks in a pond: 4 flew away: how many were left?

4. There are 8 beans in a row: if you take 4 of them away, how many will remain?

5. A window contained 9 panes of glass: a boy broke 5: how many were not broken?

6. There were 10 trees standing in a field: a storm blew down 4: how many remained?

7. Oliver is 4 years old, and Jane is 11: how much older is Jane than Oliver?

8. If 8 oranges be taken from 12 oranges, how many will remain?

9. A hen had 13 chickens, and 4 of them died: how many remained alive?

LESSON XXIX.

5 from 5 leaves 0	5 from 10 leaves 5
1 from 6 leaves 5	6 from 11 leaves 5
5 from 6 leaves 1	5 from 11 leaves 6
2 from 7 leaves 5	7 from 12 leaves 5
5 from 7 leaves 2	5 from 12 leaves 7
3 from 8 leaves 5	8 from 13 leaves 5
5 from 8 leaves 3	5 from 13 leaves 8
4 from 9 leaves 5	9 from 14 leaves 5
5 from 9 leaves 4	5 from 14 leaves 9

1. Francis had 6 oranges, and gave 5 of them away: how many had he left?

SOLUTION.—Francis had left 6 oranges less 5 oranges, which is 1 orange.

2. There are 7 crows in the field: if 2 of them fly away, how many will be left?

3. Of 8 ships that went to sea, 5 were lost in a storm: how many remained?

4. Lucy had 9 yards of ribbon, and gave 5 of them for a doll: how many yards had she left?

5. Daniel, having 10 marbles, gave 5 of them to John: how many had he left?

6. Father gave me 6 cents, and mother gave me enough more to make 11 cents: how many did she give me?

7. I owe 12 cents, and have but 5: how many more must I get to pay the debt?

8. I owed 13 dollars, and paid all but 5 dollars: how much did I pay?

9. Mary had 14 nuts: she cracked and ate 9 of them: how many nuts has she left?

LESSON XXX.

6 from 6 leaves 0	5 from 11 leaves 6
1 from 7 leaves 6	6 from 11 leaves 5
6 from 7 leaves 1	6 from 12 leaves 6
2 from 8 leaves 6	7 from 13 leaves 6
6 from 8 leaves 2	6 from 13 leaves 7
3 from 9 leaves 6	8 from 14 leaves 6
6 from 9 leaves 3	6 from 14 leaves 8
4 from 10 leaves 6	9 from 15 leaves 6
6 from 10 leaves 4	6 from 15 leaves 9

1. If James makes 7 marks on a slate, and then rubs out 1 of them, how many remain?

SOLUTION.—There remain 7 marks less 1 mark, which are 6 marks.

2. Mary had 8 cents, and spent 6 cents for a thimble: how much had she left?

3. Henry had 9 raisins: after eating 6 of them, how many had he left?

4. I bought 10 eggs in market, and broke 4 of them coming home: how many whole eggs remained?

5. If 6 cents be taken from 11 cents, how many cents will remain?

6. I paid 6 dollars for a coat, and sold it for 12 dollars: how much did I gain?

7. I had 13 dozen eggs, and sold 6 dozen: how many dozen were left?

8. John had 14 oranges, and gave away 8 of them: how many oranges had he then?

9. Nine and how many make 15? Six and how many make 14?

LESSON XXXI.

7 from 7 leaves 0	5 from 12 leaves 7
1 from 8 leaves 7	7 from 12 leaves 5
7 from 8 leaves 1	6 from 13 leaves 7
2 from 9 leaves 7	7 from 13 leaves 6
7 from 9 leaves 2	7 from 14 leaves 7
3 from 10 leaves 7	8 from 15 leaves 7
7 from 10 leaves 3	7 from 15 leaves 8
4 from 11 leaves 7	9 from 16 leaves 7
7 from 11 leaves 4	7 from 16 leaves 9

1. Charles had 8 cents, and spent 7 cents for candy: how much had he left?

SOLUTION.—Charles had left 8 cents less 7 cents, which is 1 cent.

2. A man bought a calf for 9 dollars, and paid 2 dollars down: how much remained unpaid?

3. Ella bought 10 oranges, and gave her mother 7: how many had she left?

4. A book, which cost 7 cents, was sold for 11 cents: how much was gained?

5. Thomas had 12 marbles, and lost 5 of them: how many had he left?

6. If 7 lemons be taken from 13 lemons, how many lemons will remain?

7. Mary is 14 years old, and Anna is 7: how much older is Mary than Anna?

8. A boy counted 15 birds on a tree: some of them flying away, he counted 8 remaining: how many flew away?

9. Seven and how many make 15? Nine and how many make 16?

LESSON XXXII.

8 from 8 leaves 0	5 from 13 leaves 8
1 from 9 leaves 8	8 from 13 leaves 5
8 from 9 leaves 1	6 from 14 leaves 8
2 from 10 leaves 8	8 from 14 leaves 6
8 from 10 leaves 2	7 from 15 leaves 8
3 from 11 leaves 8	8 from 15 leaves 7
8 from 11 leaves 3	8 from 16 leaves 8
4 from 12 leaves 8	9 from 17 leaves 8
8 from 12 leaves 4	8 from 17 leaves 9

1. Mary has 9 pecans: after eating 1 of them, how many will she have left?

SOLUTION.—Mary will have left 9 pecans less 1 pecan, which are 8 pecans.

2. Jane has 10 cents: if she give 8 cents for a book, how many cents will she still have?

3. Susan bought 11 peaches, and gave Emma 8: how many had she left?

4. I sold a ball for 12 cents, which cost me 8 cents: how much did I gain?

5. Frank had 13 oranges, and gave 5 to Charles: how many did Frank then have?

6. Samuel, having 14 marbles, lost 8 of them: how many marbles had he left?

7. Henry bought 15 pens, and lost 7: how many pens had he remaining?

8. If 16 persons are in a room, and 8 of them leave, how many remain?

9. Nine and how many make 17? Eight and how many make 15?

LESSON XXXIII.

9 from 9 leaves 0	5 from 14 leaves 9
1 from 10 leaves 9	9 from 14 leaves 5
9 from 10 leaves 1	6 from 15 leaves 9
2 from 11 leaves 9	9 from 15 leaves 6
9 from 11 leaves 2	7 from 16 leaves 9
3 from 12 leaves 9	9 from 16 leaves 7
9 from 12 leaves 3	8 from 17 leaves 9
4 from 13 leaves 9	9 from 17 leaves 8
9 from 13 leaves 4	9 from 18 leaves 9

1. Thomas has 10 walnuts: after eating 9, how many will he have left?

SOLUTION.—Thomas will have left 10 walnuts less 9 walnuts, which is 1 walnut.

2. I had 11 cents, and spent 9 for a slate: how much had I remaining?

3. Having 12 dollars, I spent 9 for a coat: how much had I then?

4. Anna had 13 birds: 4 of them died: how many had she left?

5. There were 14 horses in a field, but 9 of them got out: how many horses remained?

6. Mary had 15 plums, and gave her brother 6: how many had she then?

7. George had 16 marbles, and lost 9 of them: how many did he then have?

8. A man had 17 horses, and sold 8 of them: how many had he then?

9. I bought a kite for 9 cents, and sold it for 18 cents: how much did I make?

LESSON XXXIV.

10 from 10 leaves 0	5 from 15 leaves 10
1 from 11 leaves 10	10 from 15 leaves 5
10 from 11 leaves 1	6 from 16 leaves 10
2 from 12 leaves 10	10 from 16 leaves 6
10 from 12 leaves 2	7 from 17 leaves 10
3 from 13 leaves 10	10 from 17 leaves 7
10 from 13 leaves 3	8 from 18 leaves 10
4 from 14 leaves 10	10 from 18 leaves 8
10 from 14 leaves 4	9 from 19 leaves 10

1. Lucy had 11 cents, and paid 10 for a book: how many cents had she left?

SOLUTION.—Lucy had left 11 cents less 10 cents, which is 1 cent.

2. I had 12 dollars: after spending part, I had 10 dollars left: how much did I spend?
3. There are 13 pupils in school: if 3 of them leave, how many will remain?
4. There were 14 trees standing in a field: 10 of them were cut down: how many remained?
5. Charles had 15 marbles, and lost all but 5: how many did he lose?
6. Henry had 16 cents, and spent 10: how many did he then have?
7. I owe 17 dollars: if I pay all but 7, how many will I pay?
8. I bought three oranges for 10 cents, and sold them for 18 cents: how much did I gain?
9. 10 from 19 leaves how many? 9 from 19 leaves how many?

LESSON XXXV.

REVIEW.

1. How many are 4 less 2?

SOLUTION.—4 less 2 are 2.

2. How many are 10 less 4? 10 less 6? 6 less 3?
3. How many are 7 less 5? 9 less 3? 11 less 4?
4. How many are 12 less 4? 12 less 8? 13 less 6?
5. How many are 18 less 10? 19 less 9? 15 less 9?
6. How many are 16 less 10? 13 less 3? 14 less 7?
7. How many are 11 less 8? 7 less 4? 11 less 6?
8. How many are 12 less 3? 16 less 7? 15 less 10?
9. How many are 17 less 8? 17 less 10? 11 less 9?
10. How many are 9 less 5? 8 less 3? 7 less 2?

11. How many are 8 less 6? 10 less 7? 11 less 5?
12. How many are 12 less 9? 13 less 10? 12 less 10?
13. How many are 14 less 9? 13 less 7? 15 less 8?
14. How many are 17 less 9? 13 less 4? 16 less 8?
15. How many are 11 less 7? 18 less 9? 15 less 7?
16. How many are 16 less 9? 13 less 5? 12 less 6?
17. How many are 14 less 10? 14 less 8? 9 less 4?
18. How many are 15 less 6? 12 less 5? 8 less 5?
19. How many are 11 less 3? 15 less 5? 12 less 7?
20. How many are 9 less 2? 19 less 10? 17 less 7?

21. How many are 5 less 2? 10 less 3? 12 less 2?
22. How many are 18 less 8? 5 less 3? 6 less 4?
23. How many are 6 less 2? 14 less 6? 9 less 6?
24. How many are 16 less 6? 14 less 4? 8 less 2?
25. How many are 8 less 4? 13 less 9? 9 less 7?
26. How many are 10 less 2? 14 less 5? 11 less 2?
27. How many are 13 less 8? 10 less 8? 10 less 5?

LESSON XXXVI.

ADDITION AND SUBTRACTION.

1. How many are 2 and 5, less 3?

SOLUTION.—2 and 5 are 7; 7 less 3 are 4.

2. How many are 3 and 3, less 4?
3. How many are 8 and 9, less 7? 9 and 8, less 10?
4. How many are 7 and 10, less 8? 10 and 7, less 9?
5. How many are 3 and 4, less 5? 5 and 2, less 4?
6. How many are 2 and 6, less 3? 5 and 3, less 4?
7. How many are 7 and 9, less 6? 10 and 6, less 8?
8. How many are 3 and 5, less 6? 4 and 4, less 6?
9. How many are 6 and 10, less 9? 10 and 6, less 7?
10. How many are 8 and 10, less 9? 9 and 9, less 10?

11. How many are 3 and 9, less 10? 9 and 3, less 7?
12. How many are 7 and 7, less 5? 8 and 5, less 6?
13. How many are 4 and 8, less 9? 5 and 9, less 7?
14. How many are 6 and 7, less 3? 7 and 8, less 5?
15. How many are 5 and 6, less 9? 2 and 7, less 4?
16. How many are 6 and 6, less 4? 10 and 5, less 7?
17. How many are 3 and 8, less 5? 8 and 7, less 6?
18. How many are 5 and 5, less 2? 8 and 6, less 4?
19. How many are 8 and 6, less 9? 7 and 5, less 8?
20. How many are 3 and 7, less 4? 6 and 8, less 10?

21. How many are 4 and 9, less 5? 9 and 5, less 8?
22. How many are 8 and 8, less 10? 6 and 4, less 3?
23. How many are 6 and 3, less 5? 8 and 6, less 7?
24. How many are 5 and 4, less 2? 2 and 8, less 5?
25. How many are 9 and 6, less 7? 7 and 6, less 10?
26. How many are 7 and 4, less 8? 10 and 2, less 5?
27. How many are 5 and 7, less 3? 6 and 3, less 7?

LESSON XXXVII.

1. Begin with 20, and subtract by 2's to 0.

SOLUTION.—20, 18, 16, 14, 12, 10, 8, 6, 4, 2, 0.

2. Begin with 18, and subtract by 3's to 0.
3. Begin with 20, and subtract by 4's to 0.
4. Begin with 20, and subtract by 5's to 0.
5. Begin with 18, and subtract by 6's to 0.
6. Francis has 10 cents in two pockets: there are 4 cents in one: how many in the other?
7. I think of two numbers that together make 8: one of them is 5: what is the other?
8. Mary had 11 apples: she gave 4 to Lucy, and 5 to Nancy: how many had she left?

SOLUTION.—Mary gave away 4 apples and 5 apples, which are 9 apples; then, she had left 11 apples less 9 apples, which are 2 apples.

9. Emma had 15 cents: she paid 5 cents for thread, 2 cents for tape, and 3 cents for needles: how much had she left?
10. I have 10 cents in one hand, and 5 in the other: if I take 3 cents from each hand, how many cents will I then have in both hands?
11. Three numbers together make 18: the first number is 9, the second is 4: what is the third?
12. I bought 16 oranges, and gave 6 to James: Henry afterwards gave me 8 more: how many oranges had I then?
13. James has 8 marbles, and John has 7; Henry has 10 less than James and John together: how many marbles has Henry?
14. Albert bought 10 apples: he sold 3, and ate 2: how many apples had he left?

LESSON XXXVIII.

How many swallows can you count in the picture? How many are flying inside the shed? How many are at rest? How many are flying outside the shed?

1. How many birds are three times four birds?
2. How many are 3 times 4?
3. There are four nests over the door-way: if there are four eggs in each nest, how many eggs are there in all the nests?
4. How many are 4 times 4?
5. Each swallow has two wings: how many wings have eight swallows?
6. How many are 8 times 2?

LESSON XXXIX.

1 time 2 is 2	6 times 1 are 6
2 times 1 are 2	1 time 7 is 7
1 time 3 is 3	7 times 1 are 7
3 times 1 are 3	1 time 8 is 8
1 time 4 is 4	8 times 1 are 8
4 times 1 are 4	1 time 9 is 9
1 time 5 is 5	9 times 1 are 9
5 times 1 are 5	1 time 10 is 10
1 time 6 is 6	10 times 1 are 10

1. John bought 2 figs at 1 cent each: how much did they cost?

SOLUTION.—They cost 2 times 1 cent, which are 2 cents.

2. Henry paid 1 cent for a slate pencil: at that rate, how much will 3 pencils cost?

3. If apples cost 1 cent apiece, what will be the cost of 4 apples?

4. If one yard of tape cost 1 cent, what will be the cost of 5 yards?

5. What will be the cost of 6 yards of silk, if one yard cost 1 dollar?

6. If a car travel 1 mile in a minute, how far will it travel in 7 minutes?

7. If one toy book cost 1 cent, how much must Mary pay for 8 toy books?

8. How much will 9 yards of tape cost, at 1 cent a yard?

9. If eggs are worth 1 cent each, what will 10 eggs cost?

LESSON XL.

2 times 1 are 2	6 times 2 are 12
2 times 2 are 4	2 times 7 are 14
2 times 3 are 6	7 times 2 are 14
3 times 2 are 6	2 times 8 are 16
2 times 4 are 8	8 times 2 are 16
4 times 2 are 8	2 times 9 are 18
2 times 5 are 10	9 times 2 are 18
5 times 2 are 10	2 times 10 are 20
2 times 6 are 12	10 times 2 are 20

1. When peaches are selling at 2 cents each, how much will 2 peaches cost?

SOLUTION.—They will cost 2 times 2 cents, which are 4 cents.

2. Frank bought 2 plums, at 3 cents each: how much did they cost?
3. How much must you pay for 4 yards of tape, at 2 cents a yard?
4. How much will 2 pounds of meat cost, at 5 cents a pound?
5. When lemons are selling at 6 cents each, how many cents will 2 lemons cost?
6. Sarah bought 2 yards of ribbon, at 7 cents a yard: how many cents did they cost?
7. William bought 8 sticks of candy, at 2 cents a stick: how many cents did they cost?
8. Harry has 9 cents, and Emma has 2 times as many: how many cents has Emma?
9. Jane bought 2 books at 10 cents each: how much did she pay for them?

LESSON XLI.

3 times 1 are 3	6 times 3 are 18
2 times 3 are 6	3 times 7 are 21
3 times 2 are 6	7 times 3 are 21
3 times 3 are 9	3 times 8 are 24
3 times 4 are 12	8 times 3 are 24
4 times 3 are 12	3 times 9 are 27
3 times 5 are 15	9 times 3 are 27
5 times 3 are 15	3 times 10 are 30
3 times 6 are 18	10 times 3 are 30

1. If a pint of chestnuts cost 3 cents, how much will 2 pints cost?

SOLUTION.—They will cost 2 times 3 cents, which are 6 cents.

2. James bought 3 tops, at 3 cents each: what did he pay for them?
3. If Thomas can walk 3 miles in an hour, how far can he walk in 4 hours?
4. If 1 pear is worth 5 apples, how many apples are 3 pears worth?
5. If 1 peach is worth 3 plums, how many plums are 6 peaches worth?
6. How much will 7 yards of tape cost, if 1 yard cost 3 cents?
7. If 1 orange cost 3 cents, how many cents will 8 oranges cost?
8. If 1 pound of sugar cost 9 cents, how much will 3 pounds cost?
9. A slate cost 10 cents, and a book 3 times as much: what did the book cost?

LESSON XLII.

4 times 1 are 4	6 times 4 are 24
2 times 4 are 8	4 times 7 are 28
4 times 2 are 8	7 times 4 are 28
3 times 4 are 12	4 times 8 are 32
4 times 3 are 12	8 times 4 are 32
4 times 4 are 16	4 times 9 are 36
4 times 5 are 20	9 times 4 are 36
5 times 4 are 20	4 times 10 are 40
4 times 6 are 24	10 times 4 are 40

1. Lucy has 2 kittens, and each one has 4 feet: how many feet have both?

SOLUTION.—Both kittens have 2 times 4 feet, which are 8 feet.

2. Thomas has 3 pigeons, and James has 4 times as many: how many has James?

3. Daniel bought 4 tops, at 4 cents each: what did they cost?

4. When rice is 4 cents a pound, how much will 5 pounds cost?

5. There are 4 quarters in one apple: how many quarters are there in 6 apples?

6. Francis bought 7 oranges, at 4 cents each: what did they cost?

7. If peaches are sold at 4 cents apiece, how much will 8 peaches cost?

8. If a pound of starch cost 9 cents, what will 4 pounds cost?

9. At 10 cents apiece, what will be the cost of 4 lead-pencils?

LESSON XLIII.

5 times 1 are 5	6 times 5 are 30
2 times 5 are 10	5 times 7 are 35
5 times 2 are 10	7 times 5 are 35
3 times 5 are 15	5 times 8 are 40
5 times 3 are 15	8 times 5 are 40
4 times 5 are 20	5 times 9 are 45
5 times 4 are 20	9 times 5 are 45
5 times 5 are 25	5 times 10 are 50
5 times 6 are 30	10 times 5 are 50

1. Francis bought 5 tops, at 2 cents each: how many cents did they cost?

SOLUTION.—Five tops cost 5 times 2 cents, which are 10 cents.

2. If James can walk 5 miles in 1 hour, how many miles can he walk in 3 hours?

3. If 1 peach is worth 5 plums, how many plums are 4 peaches worth?

4. Lucy has 5 hens, and each hen has 5 chickens: how many chickens have all the hens?

5. Francis found 6 bird's nests, with 5 eggs in each: how many eggs in all?

6. Daniel bought 5 oranges, at 7 cents each: how much did he pay for them?

7. What will be the cost of 8 kites, at 5 cents apiece?

8. If 1 pound of flour cost 5 cents, what will 9 pounds cost?

9. If one street-car ticket cost 5 cents, what will 10 tickets cost?

LESSON XLIV.

6 times 1 are 6	6 times 6 are 36
2 times 6 are 12	6 times 7 are 42
6 times 2 are 12	7 times 6 are 42
3 times 6 are 18	6 times 8 are 48
6 times 3 are 18	8 times 6 are 48
4 times 6 are 24	6 times 9 are 54
6 times 4 are 24	9 times 6 are 54
5 times 6 are 30	6 times 10 are 60
6 times 5 are 30	10 times 6 are 60

1. If 1 dress can be made from 6 yards of calico, how many yards will it take to make 2 dresses?

SOLUTION.—For 2 dresses it will take 2 times 6 yards, which are 12 yards.

2. Mary has 3 hens, and each hen has 6 chickens: how many chickens are there in all?

3. If there are 6 panes of glass in one window, how many panes are there in 4 windows?

4. If 1 orange is worth 5 peaches, how many peaches are 6 oranges worth?

5. John bought 6 quarts of plums, at 6 cents a quart: how much did they cost?

6. If 1 quart of strawberries cost 7 cents, how much will 6 quarts cost?

7. If a man eat 6 ounces of bread in 1 day, how many ounces will he eat in 8 days?

8. What will be the cost of 6 lead-pencils, at 9 cents each?

9. I bought 6 dozen eggs, at 10 cents a dozen: how much did they cost?

LESSON XLV.

7 times 1 are 7	6 times 7 are 42
2 times 7 are 14	7 times 6 are 42
7 times 2 are 14	7 times 7 are 49
3 times 7 are 21	7 times 8 are 56
7 times 3 are 21	8 times 7 are 56
4 times 7 are 28	7 times 9 are 63
7 times 4 are 28	9 times 7 are 63
5 times 7 are 35	7 times 10 are 70
7 times 5 are 35	10 times 7 are 70

1. Sarah bought 2 thimbles, at 7 cents each: how much did both cost?

SOLUTION.—Both thimbles cost 2 times 7 cents, which are 14 cents.

2. Edward has 3 pockets, and has 7 marbles in each: how many marbles has he?

3. There are 7 days in a week: how many days are there in 4 weeks?

4. If one melon is worth 5 peaches, how many peaches are 7 melons worth?

5. If a horse travel 7 miles in one hour, how many miles will he travel in 6 hours?

6. If each of 7 benches will seat 7 boys, how many boys can be seated on them all?

7. If Harry gives 7 marbles for one cent, how many must he give for 8 cents?

8. If muslin is 9 cents a yard, how many cents will 7 yards cost?

9. At 10 cents a yard, how much will 7 yards of ribbon cost?

LESSON XLVI.

8 times 1 are 8	6 times 8 are 48
2 times 8 are 16	8 times 6 are 48
8 times 2 are 16	7 times 8 are 56
3 times 8 are 24	8 times 7 are 56
8 times 3 are 24	8 times 8 are 64
4 times 8 are 32	8 times 9 are 72
8 times 4 are 32	9 times 8 are 72
5 times 8 are 40	8 times 10 are 80
8 times 5 are 40	10 times 8 are 80

1. James bought 2 melons, at 8 cents each: how many cents did they cost?

Solution.—The melons cost 2 times 8 cents, which are 16 cents.

2. Each one of 3 boys caught 8 fishes, how many did they all catch?
3. Nancy has 4 hens, and each hen has 8 chickens: how many chickens are there?
4. There are 5 houses, each having 8 windows: how many windows in all the houses?
5. If there are 8 pints in one gallon, how many pints are there in 6 gallons?
6. Clara bought 8 spools of thread, at 7 cents each: how much did they cost?
7. There are 8 quarts in 1 peck: how many quarts are there in 8 pecks?
8. If one dozen apples cost 9 cents, how many cents will 8 dozen apples cost?
9. If one comb cost 10 cents, what will be the cost of 8 combs?

LESSON XLVII.

9 times 1 are 9	6 times 9 are 54
2 times 9 are 18	9 times 6 are 54
9 times 2 are 18	7 times 9 are 63
3 times 9 are 27	9 times 7 are 63
9 times 3 are 27	8 times 9 are 72
4 times 9 are 36	9 times 8 are 72
9 times 4 are 36	9 times 9 are 81
5 times 9 are 45	9 times 10 are 90
9 times 5 are 45	10 times 9 are 90

1. Francis bought 2 knives, at 9 cents each: how many cents did they cost?

SOLUTION.—The knives cost 2 times 9 cents, which are 18 cents.

2. I bought 3 pounds of raisins, at 9 cents a pound: what did I pay for them?

3. There are 9 panes of glass in each of 4 windows: how many panes in all?

4. If one orange cost 5 cents, how much will 9 oranges cost?

5. How many cents will I have to pay for 6 tops, at 9 cents each?

6. Frank bought 9 lemons, at 7 cents each: how much did they cost?

7. If a boy travel 8 miles in 1 hour, how far will he travel in 9 hours?

8. How many cents must be paid for 9 yards of muslin, at 9 cents a yard?

9. In one dime there are 10 cents: how many cents in 9 dimes?

LESSON XLVIII.

10 times 1 are 10	6 times 10 are 60
2 times 10 are 20	10 times 6 are 60
10 times 2 are 20	7 times 10 are 70
3 times 10 are 30	10 times 7 are 70
10 times 3 are 30	8 times 10 are 80
4 times 10 are 40	10 times 8 are 80
10 times 4 are 40	9 times 10 are 90
5 times 10 are 50	10 times 9 are 90
10 times 5 are 50	10 times 10 are 100

1. I bought 10 pencils, at 2 cents each: how much did they cost?

SOLUTION.—The pencils cost 10 times 2 cents, which are 20 cents.

2. If George earn 3 dollars in 1 week, how much will he earn in 10 weeks?

3. There are 4 pecks in 1 bushel: how many pecks are there in 10 bushels?

4. If a man can eat 5 pounds of bread in 1 week, how much can he eat in 10 weeks?

5. At 6 dollars a cord, what will 10 cords of wood cost?

6. If 10 marbles are given for 1 cent, how many must be given for 7 cents?

7. How many cents will pay for 10 oranges, at 8 cents each?

8. What will 9 barrels of flour cost, at 10 dollars a barrel?

9. At 10 dollars a yard, how much will 10 yards of cloth cost?

LESSON XLIX.

REVIEW.

1. How many are 2 times 5?

SOLUTION.—2 times 5 are 10.

2. How many are 3 times 4? 2 times 9?
3. How many are 10 times 2? 3 times 3?
4. How many are 4 times 2? 5 times 4?
5. How many are 8 times 2? 2 times 2?
6. How many are 3 times 10? 5 times 6?
7. How many are 6 times 2? 8 times 3?
8. How many are 4 times 10? 7 times 3?
9. How many are 7 times 4? 5 times 5?
10. How many are 3 times 5? 6 times 3?

11. How many are 4 times 9? 6 times 4?
12. How many are 9 times 3? 4 times 8?
13. How many are 4 times 4? 9 times 5?
14. How many are 5 times 7? 10 times 8?
15. How many are 8 times 6? 7 times 10?
16. How many are 6 times 6? 8 times 8?
17. How many are 9 times 7? 6 times 9?
18. How many are 5 times 8? 7 times 7?
19. How many are 10 times 5? 6 times 10?
20. How many are 7 times 6? 9 times 10?

21. How many are 8 times 7? 9 times 9?
22. How many are 8 times 9? 10 times 10?
23. How many are 3 times 2? 7 times 2?
24. How many are 5 times 3? 6 times 7?
25. How many are 3 times 8? 9 times 6?
26. How many are 8 times 5? 7 times 9?
27. How many are 10 times 9? 9 times 8?

28. How many are 2 times 3 times 3?

SOLUTION.—3 times 3 are 9; 2 times 9 are 18.

29. How many are 4 times 2 times 2?
30. How many are 2 times 3 times 4?
31. How many are 2 times 2 times 5?
32. How many are 5 times 2 times 3?
33. How many are 3 times 2 times 6?
34. How many are 2 times 2 times 6?
35. How many are 2 times 3 times 7?
36. How many are 2 times 2 times 7?
37. How many are 4 times 2 times 8?
38. How many are 3 times 3 times 4?
39. How many are 2 times 4 times 6?
40. How many are 2 times 2 times 10?

41. How many are 3 times 2 times 10?
42. How many are 4 times 2 times 5?
43. How many are 4 times 2 times 4?
44. How many are 3 times 2 times 8?
45. How many are 4 times 2 times 10?
46. How many are 2 times 5 times 7?
47. How many are 3 times 3 times 10?
48. How many are 5 times 2 times 8?
49. How many are 2 times 5 times 9?
50. How many are 5 times 2 times 6?

51. How many are 2 times 3 times 9?
52. How many are 2 times 2 times 9?
53. How many are 3 times 3 times 5?
54. How many are 3 times 3 times 6?
55. How many are 2 times 5 times 10?
56. How many are 4 times 2 times 9?
57. How many are 3 times 3 times 7?
58. How many are 3 times 3 times 8?

LESSON L.

REVIEW.

1. How many are 2 and 5, less 4, multiplied by 3?

SOLUTION.—2 and 5 are 7; 7 less 4 are 3; 3 multiplied by 3 is 9.

2. How many are 3 and 5, less 4, multiplied by 2?
3. How many are 4 and 6, less 5, multiplied by 10?
4. How many are 5 and 5, less 4, multiplied by 7?
5. How many are 6 and 5, less 4, multiplied by 7?
6. How many are 7 and 6, less 5, multiplied by 4?
7. How many are 8 and 7, less 6, multiplied by 3?
8. How many are 2 and 6, less 5, multiplied by 4?
9. How many are 3 and 6, less 5, multiplied by 7?

10. How many are 4 and 7, less 6, multiplied by 5?
11. How many are 5 and 7, less 6, multiplied by 8?
12. How many are 6 and 6, less 5, multiplied by 2?
13. How many are 7 and 7, less 6, multiplied by 8?
14. How many are 8 and 8, less 7, multiplied by 9?
15. How many are 2 and 7, less 6, multiplied by 5?
16. How many are 4 and 8, less 7, multiplied by 9?
17. How many are 3 and 7, less 6, multiplied by 5?
18. How many are 5 and 8, less 7, multiplied by 9?

19. How many are 2 and 8, less 7, multiplied by 6?
20. How many are 3 and 9, less 8, multiplied by 6?
21. How many are 7 and 10, less 9, multiplied by 2?
22. How many are 2 and 9, less 8, multiplied by 7?
23. How many are 3 and 8, less 7, multiplied by 9?
24. How many are 4 and 10, less 9, multiplied by 8?
25. How many are 6 and 10, less 9, multiplied by 8?
26. How many are 6 and 8, less 7, multiplied by 9?
27. How many are 4 and 9, less 8, multiplied by 6?

LESSON LI.

PROMISCUOUS QUESTIONS.

1. Joseph had 14 cents, and bought 2 oranges, at 5 cents each: how much money had he left?

SOLUTION.—Joseph paid for the oranges 2 times 5 cents, which are 10 cents; he had left 14 cents, less 10 cents, which are 4 cents.

2. James bought a calf for 8 dollars, and 3 sheep, at 4 dollars apiece: how much did he pay for all?

3. George owed me 19 cents: he gave me 2 oranges, worth 5 cents each, and the remainder in money: how much money did I get?

4. I bought 2 yards of cloth, at 4 dollars a yard, and 3 yards, at 2 dollars a yard: how much did all cost?

5. A boy worked 4 weeks, at 4 dollars a week: he spent 5 dollars for a coat, and 2 dollars for a hat: how much money had he left?

6. A man can dig 6 bushels of potatoes in one hour, and a boy 3 bushels: how many bushels can both dig in 8 hours?

7. A farmer has 9 hogs: 4 of them die, and he sells the rest at 10 dollars apiece: how much did he get for them?

8. Mary had 8 five-cent pieces, and lost 5 of them: how many cents had she left?

9. A blacksmith shod 7 horses in one day, putting a shoe on each foot: how many shoes did he use?

10. Kate received 10 cents one day, and 8 cents another: she then spent 4 cents for apples and 5 cents for candy, after which her father gave her six times as much as she had left: how much did her father give her?

LESSON LII.

In the picture you can count five wild geese and twenty ducks.

1. If twenty ducks were divided into four flocks, each having the same number, how many ducks would there be in each flock?

2. How many times is 5 contained in 20?

3. If five wild geese have ten wings, how many wings has each goose?

4. How many times is 5 contained in 10?

5. There are four full-blown flowers on two branches: how many on each branch?

LESSON LIII.

2 in 2, 1 time	6 in 12, 2 times
2 in 4, 2 times	2 in 14, 7 times
2 in 6, 3 times	7 in 14, 2 times
3 in 6, 2 times	2 in 16, 8 times
2 in 8, 4 times	8 in 16, 2 times
4 in 8, 2 times	2 in 18, 9 times
2 in 10, 5 times	9 in 18, 2 times
5 in 10, 2 times	2 in 20, 10 times
2 in 12, 6 times	10 in 20, 2 times

1. How many apples, at 2 cents each, can you buy for 4 cents?

SOLUTION.—You can buy as many apples as 2 cents are contained times in 4 cents, which are 2 times. Hence you can buy 2 apples.

2. How many marbles, at 2 cents each, can you buy for 6 cents?

3. How many lemons, at 4 cents each, can you buy for 8 cents?

4. How many peaches, at 5 cents each, can you buy for 10 cents?

5. How many yards of ribbon, at 2 cents a yard, can you buy for 12 cents?

6. How many oranges, at 7 cents each, can you buy for 14 cents?

7. How many tops, at 2 cents each, can you buy for 16 cents?

8. How many kites, at 9 cents each, can you buy for 18 cents?

9. How many books, at 10 cents apiece, can you buy for 20 cents?

LESSON LIV.

3 in 3, 1 time	6 in 18, 3 times
3 in 6, 2 times	3 in 21, 7 times
2 in 6, 3 times	7 in 21, 3 times
3 in 9, 3 times	3 in 24, 8 times
3 in 12, 4 times	8 in 24, 3 times
4 in 12, 3 times	3 in 27, 9 times
3 in 15, 5 times	9 in 27, 3 times
5 in 15, 3 times	3 in 30, 10 times
3 in 18, 6 times	10 in 30, 3 times

1. If you have 6 balls, how many groups, of 3 balls each, can you make out of them?

SOLUTION.—You can make as many groups as 3 balls are contained times in 6 balls, which are 2 times. Hence you can make 2 groups.

2. Jane paid 9 cents for ribbon, at 3 cents a yard: how many yards did she get?

3. When pears are 4 cents each, how many can you buy for 12 cents?

4. In one yard there are 3 feet: how many yards are there in 15 feet?

5. If one orange is worth 3 lemons, how many oranges can you get for 18 lemons?

6. I have 21 marbles in groups of 7 each: how many groups are there?

7. How much cloth, at 8 dollars a yard, can you buy for 24 dollars?

8. For 27 cents, I bought 9 peaches: how much did one peach cost?

9. How many 3-cent postage-stamps can you buy for 30 cents?

LESSON LV.

4 in 4, 1 time	6 in 24, 4 times
4 in 8, 2 times	4 in 28, 7 times
2 in 8, 4 times	7 in 28, 4 times
4 in 12, 3 times	4 in 32, 8 times
3 in 12, 4 times	8 in 32, 4 times
4 in 16, 4 times	4 in 36, 9 times
4 in 20, 5 times	9 in 36, 4 times
5 in 20, 4 times	4 in 40, 10 times
4 in 24, 6 times	10 in 40, 4 times

1. How many oranges, at 4 cents each, can you buy for 8 cents?

SOLUTION.—You can buy as many oranges as 4 cents are contained times in 8 cents, which are 2 times. Hence you can buy 2 oranges.

2. There are 4 quarts in one gallon: how many gallons are there in 12 quarts?

3. If 16 apples be divided equally among 4 boys, how many will each have?

4. There are 20 scholars sitting on 4 benches: how many scholars on each bench?

5. If 4 sheets of paper make one copy-book, how many copy-books will 24 sheets make?

6. If one top cost 7 cents, how many tops can be bought for 28 cents?

7. At 4 cents each, how many peaches can you buy for 32 cents?

8. At 9 cents each, how many cakes can you buy for 36 cents?

9. If a spelling-book cost 10 cents, how many spelling-books can you buy for 40 cents?

LESSON LVI.

5 in 5, 1 time	6 in 30, 5 times
5 in 10, 2 times	5 in 35, 7 times
2 in 10, 5 times	7 in 35, 5 times
5 in 15, 3 times	5 in 40, 8 times
3 in 15, 5 times	8 in 40, 5 times
5 in 20, 4 times	5 in 45, 9 times
4 in 20, 5 times	9 in 45, 5 times
5 in 25, 5 times	5 in 50, 10 times
5 in 30, 6 times	10 in 50, 5 times

1. How many oranges, at 5 cents each, can you buy for 10 cents?

SOLUTION.—You can buy as many oranges as 5 cents are contained times in 10 cents, which are 2 times. Hence you can buy 2 oranges.

2. How many pencils, at 3 cents each, can you buy for 15 cents?

3. How many toy-books, at 4 cents each, can you buy for 20 cents?

4. How many pears, at 5 cents each, can you buy for 25 cents?

5. How many melons, at 6 cents each, can you buy for 30 cents?

6. In one week there are 7 days: how many weeks in 35 days?

7. How many cakes, at 8 cents each, can you buy for 40 cents?

8. How many tops, at 5 cents each, can you buy for 45 cents?

9. How many slates, at 5 cents each, can be bought for 50 cents?

LESSON LVII.

6 in 6, 1 time	6 in 36, 6 times
6 in 12, 2 times	6 in 42, 7 times
2 in 12, 6 times	7 in 42, 6 times
6 in 18, 3 times	6 in 48, 8 times
3 in 18, 6 times	8 in 48, 6 times
6 in 24, 4 times	6 in 54, 9 times
4 in 24, 6 times	9 in 54, 6 times
6 in 30, 5 times	6 in 60, 10 times
5 in 30, 6 times	10 in 60, 6 times

1. How many quarts of milk, at 6 cents a quart, can you buy for 12 cents?

SOLUTION.—You can buy as many quarts as 6 cents are contained times in 12 cents, which are 2 times. Hence you can buy 2 quarts.

2. How many oranges, at 6 cents each, can you buy for 18 cents?

3. There are 24 trees in 6 rows: how many trees in each row?

4. How many pears, at 5 cents each, can you buy for 30 cents?

5. How many pounds of flour, at 6 cents a pound, can be bought for 36 cents?

6. How many lemons, at 7 cents each, can you buy for 42 cents?

7. How many pencils, at 8 cents each, can you buy for 48 cents?

8. How many rings, at 6 dimes each, can you buy for 54 dimes?

9. A farmer sold 6 loads of hay for 60 dollars: what did he get for each load?

LESSON LVIII.

7 in 7, 1 time	7 in 42, 6 times
7 in 14, 2 times	6 in 42, 7 times
2 in 14, 7 times	7 in 49, 7 times
7 in 21, 3 times	7 in 56, 8 times
3 in 21, 7 times	8 in 56, 7 times
7 in 28, 4 times	7 in 63, 9 times
4 in 28, 7 times	9 in 63, 7 times
7 in 35, 5 times	7 in 70, 10 times
5 in 35, 7 times	10 in 70, 7 times

1. If you divide 14 apples into piles, containing 7 apples each, how many piles will there be?

SOLUTION.—There will be as many piles as 7 apples are contained times in 14 apples, which are 2 times. Hence there will be 2 piles.

2. If a pine-apple cost 7 cents, how many pine-apples can you buy for 21 cents?

3. At 7 cents each, how many melons can be bought for 28 cents?

4. If 7 boys share 35 peaches equally, how many will each boy receive?

5. A man traveled 42 miles in 6 hours: how far did he travel in 1 hour?

6. I paid 49 cents for 7 quarts of strawberries: how much did I pay a quart?

7. There are 56 trees in 8 rows: how many trees are there in one row?

8. At 9 cents a yard, how many yards of muslin can you buy for 63 cents?

9. I paid 70 dollars for 7 calves: how much did I pay for each calf?

LESSON LIX.

8 in 8, 1 time	8 in 48, 6 times
8 in 16, 2 times	6 in 48, 8 times
2 in 16, 8 times	8 in 56, 7 times
8 in 24, 3 times	7 in 56, 8 times
3 in 24, 8 times	8 in 64, 8 times
8 in 32, 4 times	8 in 72, 9 times
4 in 32, 8 times	9 in 72, 8 times
8 in 40, 5 times	8 in 80, 10 times
5 in 40, 8 times	10 in 80, 8 times

1. If in one peck there are 8 quarts, how many pecks are there in 16 quarts?

SOLUTION.—There are as many pecks as 8 quarts are contained times in 16 quarts, which are 2 times. Hence there are 2 pecks.

2. If one orange is worth 8 apples, how many oranges can you get for 24 apples?
3. How many pencils, at 4 cents each, can you buy for 32 cents?
4. If 5 yards of calico cost 40 cents, how much will one yard cost?
5. If one cane cost 8 dimes, how many canes can be bought for 48 dimes?
6. At 7 cents each, how many tops can you buy for 56 cents?
7. If one peach is worth 8 plums, how many peaches are 64 plums worth?
8. Harry paid 72 cents for 9 pears: how much did he pay for each?
9. I gave 80 cents for 8 toy-books: what did they cost apiece?

LESSON LX.

9 in 9, 1 time	9 in 54, 6 times
9 in 18, 2 times	6 in 54, 9 times
2 in 18, 9 times	9 in 63, 7 times
9 in 27, 3 times	7 in 63, 9 times
3 in 27, 9 times	9 in 72, 8 times
9 in 36, 4 times	8 in 72, 9 times
4 in 36, 9 times	9 in 81, 9 times
9 in 45, 5 times	9 in 90, 10 times
5 in 45, 9 times	10 in 90, 9 times

1. At 9 cents each, how many pencils can you buy for 18 cents?

SOLUTION.—You can buy as many pencils as 9 cents are contained times in 18 cents, which are 2 times. Hence you can buy 2 pencils.

2. If 3 pounds of meat cost 27 cents, how much will one pound cost?

3. Melons were sold at the rate of 4 for 36 cents: how much was that apiece?

4. A father divided 45 cents equally among his 5 children: how much did each child get?

5. Mary gave 54 cents for 9 spools of thread: how much did she give for each spool?

6. A boy rode 63 miles in 7 hours: how many miles did he ride in one hour?

7. At 9 cents a yard, how many yards of ribbon can you buy for 72 cents?

8. If 81 blocks be placed in 9 rows, how many blocks will there be in each row?

9. How many books, at 10 cents each, can you buy for 90 cents?

LESSON LXI.

10 in 10, 1 time	10 in 60, 6 times
10 in 20, 2 times	6 in 60, 10 times
2 in 20, 10 times	10 in 70, 7 times
10 in 30, 3 times	7 in 70, 10 times
3 in 30, 10 times	10 in 80, 8 times
10 in 40, 4 times	8 in 80, 10 times
4 in 40, 10 times	10 in 90, 9 times
10 in 50, 5 times	9 in 90, 10 times
5 in 50, 10 times	10 in 100, 10 times

1. How many melons, at 10 cents each, can you buy for 20 cents?

SOLUTION.—You can buy as many melons as 10 cents are contained times in 20 cents, which are 2 times. Hence you can buy 2 melons.

2. If a quince is worth 10 apples, how many quinces can you get for 30 apples?

3. At 4 cents each, how many pears can you buy for 40 cents?

4. At 10 cents each, how many oranges can be bought for 50 cents?

5. I paid 60 cents for eggs, at 10 cents a dozen: how many dozens did I get?

6. How many coats, at 10 dollars each, can be bought for 70 dollars?

7. How many kites, at 8 cents each, can I buy for 80 cents?

8. How many balls, at 10 cents each, can be bought for 90 cents?

9. In one dime there are 10 cents: how many dimes are there in 100 cents?

LESSON LXII.

REVIEW.

State how many times the first numbers in the following table are contained in the second.

2 in 4*	4 in 12	4 in 32	5 in 35	9 in 54
2 in 8	3 in 18	5 in 30	6 in 42	7 in 63
3 in 6	2 in 20	6 in 18	4 in 40	8 in 64
5 in 10	10 in 30	7 in 28	8 in 40	9 in 63
3 in 27	6 in 12	8 in 32	10 in 50	10 in 40
5 in 20	9 in 27	8 in 16	7 in 42	6 in 60
10 in 70	4 in 16	3 in 30	2 in 18	4 in 20
2 in 10	6 in 24	4 in 24	5 in 45	9 in 36
10 in 20	7 in 21	5 in 15	10 in 60	6 in 48
3 in 9	3 in 21	9 in 18	8 in 48	8 in 72
6 in 36	10 in 80	8 in 24	10 in 90	9 in 72
4 in 8	2 in 16	7 in 35	7 in 49	7 in 70
2 in 12	5 in 25	6 in 30	5 in 50	9 in 81
3 in 15	4 in 28	5 in 40	6 in 54	8 in 80
2 in 14	3 in 24	9 in 45	9 in 90	8 in 56
3 in 12	7 in 14	4 in 36	7 in 56	10 in 100

*Solution.—2 is contained in 4 two times.

1. How many are 10 and 4, less 8, multiplied by 3, divided by 9?

Solution.—10 and 4 are 14; 14 less 8 are 6; 6 multiplied by 3 is 18; 18 divided by 9 is 2.

2. How many are 9 and 9, less 10, multiplied by 3, divided by 6?

3. How many are 3 and 6, less 7, multiplied by 8, divided by 4?
4. How many are 2 and 10, less 4, multiplied by 7, divided by 8?
5. How many are 10 and 6, less 9, multiplied by 5, divided by 7?
6. How many are 3 and 5, less 2, multiplied by 5, divided by 6?
7. How many are 2 and 5, less 3, multiplied by 9, divided by 6?
8. How many are 10 and 7, less 8, multiplied by 6, divided by 9?
9. How many are 2 and 4, less 3, multiplied by 7, divided by 3?
10. How many are 4 and 7, less 6, multiplied by 8, divided by 5?
11. How many are 6 and 4, less 3, multiplied by 9, divided by 7?
12. How many are 3 and 10, less 8, multiplied by 9, divided by 5?
13. How many are 6 and 9, less 10, multiplied by 10, divided by 5?
14. How many are 10 and 8, less 9, multiplied by 10, divided by 9?
15. How many are 8 and 7, less 6, multiplied by 4, divided by 6?
16. How many are 5 and 9, less 8, multiplied by 5, divided by 10?
17. How many are 3 and 10, less 5, multiplied by 4, divided by 8?
18. How many are 6 and 3, less 4, multiplied by 8, divided by 4?
19. How many are 9 and 5, less 10, multiplied by 6, divided by 8?

LESSON LXIII.

1. If 4 pens cost 8 cents, how much will one pen cost? If one pen cost 2 cents, how much will 6 pens cost?

2. If 6 peaches cost 24 cents, how much will one peach cost? If one peach cost 4 cents, how much will 9 peaches cost?

3. If 4 dozen eggs cost 20 cents, how much will 6 dozen cost?

SOLUTION.—If 4 dozen eggs cost 20 cents, 1 dozen eggs will cost as many cents as 4 is contained times in 20, which are 5; and 6 dozen will cost 6 times 5 cents, which are 30 cents.

4. If 3 lead-pencils cost 18 cents, how many cents will 5 pencils cost?

5. If 4 pine-apples cost 32 cents, how many cents will 5 pine-apples cost?

6. If 3 quarts of berries cost 21 cents, how many cents will 7 quarts cost?

7. What will 63 marbles cost, if 14 marbles cost 2 cents?

8. Henry spends 32 cents for pears, at 4 cents apiece, and 20 cents for pears, at 5 cents apiece: how many pears does he buy?

9. Sarah spends 50 cents for lemons, at 5 cents apiece, and gives 6 of them to her sister: how many lemons has she left?

10. Mary bought 7 cents worth of thread, and 10 cents worth of needles, giving the clerk 25 cents: how much change did she get?

11. I bought 4 pencils, at 3 cents each, and gave the clerk 3 five-cent pieces; how many cents should I receive in change?

LESSON LXIV.

The sign $+$ is read *plus*, and is placed between two numbers to show that they are to be added together.

The sign $-$ is read *minus*, and is placed between two numbers to show that the number on the right is to be taken from the number on the left.

The sign $\times$ is read *multiplied by*, and is placed between two numbers to show that the number on the left is to be multiplied by the number on the right.

The sign $\div$ is read *divided by*, and is placed between two numbers to show that the number on the left is to be divided by the number on the right.

The sign $=$ is read *equals*, or *is equal to*, and, when placed between two numbers, shows that they are equal.

1. $11 + 1 = ?$

SOLUTION.—11 plus 1 equals 12.

2. $2 + 11 = ?$ $11 + 3 = ?$ $4 + 11 = ?$ $11 + 5 = ?$ $6 + 11 = ?$ $11 + 7 = ?$ $8 + 11 = ?$ $11 + 9 = ?$ $10 + 11 = ?$
3. $12 + 1 = ?$ $2 + 12 = ?$ $12 + 3 = ?$ $4 + 12 = ?$ $12 + 5 = ?$ $6 + 12 = ?$ $12 + 7 = ?$ $8 + 12 = ?$ $12 + 9 = ?$ $10 + 12 = ?$
4. $13 + 1 = ?$ $2 + 13 = ?$ $13 + 3 = ?$ $4 + 13 = ?$ $13 + 5 = ?$ $6 + 13 = ?$ $13 + 7 = ?$ $8 + 13 = ?$ $13 + 9 = ?$ $10 + 13 = ?$
5. $14 + 1 = ?$ $2 + 14 = ?$ $14 + 3 = ?$ $4 + 14 = ?$ $14 + 5 = ?$ $6 + 14 = ?$ $14 + 7 = ?$ $8 + 14 = ?$ $14 + 9 = ?$ $10 + 14 = ?$
6. $15 + 1 = ?$ $2 + 15 = ?$ $15 + 3 = ?$ $4 + 15 = ?$ $15 + 5 = ?$ $6 + 15 = ?$ $15 + 7 = ?$ $8 + 15 = ?$ $15 + 9 = ?$ $10 + 15 = ?$

LESSON LXV.

1. How many are 16 + 1? 2 + 16? 16 + 3? 4 + 16? 16 + 5? 6 + 16? 16 + 7? 8 + 16? 16 + 9? 10 + 16?
2. How many are 17 + 1? 2 + 17? 17 + 3? 4 + 17? 17 + 5? 6 + 17? 17 + 7? 8 + 17? 17 + 9? 10 + 17?
3. How many are 18 + 1? 2 + 18? 18 + 3? 4 + 18? 18 + 5? 6 + 18? 18 + 7? 8 + 18? 18 + 9? 10 + 18?
4. How many are 19 + 1? 2 + 19? 19 + 3? 4 + 19? 19 + 5? 6 + 19? 19 + 7? 8 + 19? 19 + 9? 10 + 19?
5. How many are 20 + 2? 4 + 30? 40 + 6? 8 + 50? 60 + 10? 3 + 70? 80 + 5? 7 + 90? 90 + 9?
6. How many are 21 + 2? 31 + 4? 6 + 41? 41 + 8? 10 + 41? 3 + 51? 61 + 5? 7 + 71? 81 + 9? 9 + 91?
7. How many are 22 + 2? 4 + 32? 42 + 6? 8 + 52? 62 + 10? 3 + 72? 82 + 5? 7 + 92? 92 + 8?
8. How many are 23 + 2? 4 + 33? 43 + 6? 8 + 53? 63 + 10? 3 + 73? 83 + 5? 7 + 93?
9. How many are 24 + 2? 4 + 34? 44 + 6? 8 + 54? 64 + 10? 3 + 74? 84 + 5? 6 + 94?
10. How many are 25 + 2? 4 + 35? 45 + 6? 8 + 55? 65 + 10? 3 + 75? 85 + 5? 5 + 95?
11. How many are 26 + 2? 4 + 36? 46 + 6? 8 + 56? 66 + 10? 3 + 76? 86 + 5? 4 + 96?
12. How many are 27 + 2? 4 + 37? 47 + 6? 8 + 57? 67 + 10? 3 + 77? 87 + 5? 3 + 97?
13. How many are 28 + 2? 4 + 38? 48 + 6? 8 + 58? 68 + 10? 3 + 78? 88 + 5? 2 + 98?

LESSON LXVI.

1. How many are 29 + 2? 4 + 39? 49 + 6? 8 + 59? 69 + 10? 3 + 79? 89 + 5? 7 + 89?
2. How many are 24 + 3? 34 + 5? 44 + 7? 54 + 9? 64 + 2? 74 + 4? 84 + 6? 85 + 8?
3. How many are 25 + 6? 3 + 35? 45 + 5? 7 + 55? 65 + 9? 2 + 75? 85 + 4? 8 + 86?
4. How many are 26 + 6? 8 + 36? 46 + 3? 5 + 56? 66 + 7? 9 + 76? 86 + 2? 4 + 87?
5. How many are 27 + 7? 5 + 37? 47 + 3? 2 + 57? 67 + 4? 6 + 77? 87 + 8? 10 + 88?
6. How many are 27 + 9? 8 + 28? 38 + 5? 9 + 48? 58 + 7? 2 + 68? 78 + 4? 6 + 88?
7. How many are 27 + 10? 8 + 29? 39 + 9? 7 + 49? 59 + 5? 6 + 69? 79 + 2? 4 + 89?
8. How many are 23 + 3? 5 + 33? 43 + 7? 9 + 53? 63 + 2? 4 + 73? 83 + 6? 6 + 93?
9. How many are 24 + 5? 2 + 34? 44 + 4? 6 + 54? 64 + 8? 9 + 74? 84 + 7? 4 + 94?
10. How many are 35 + 5? 10 + 45? 55 + 9? 8 + 65? 75 + 7? 6 + 85? 95 + 3? 2 + 96?
11. How many are 26 + 8? 6 + 36? 46 + 4? 10 + 56? 66 + 8? 7 + 76? 86 + 6? 4 + 95?
12. How many are 27 + 5? 3 + 37? 47 + 7? 5 + 57? 67 + 9? 7 + 77? 87 + 10? 92 + 6?
13. How many are 28 + 7? 9 + 38? 48 + 8? 6 + 58? 68 + 9? 5 + 78? 88 + 7? 9 + 88?
14. How many are 29 + 9? 7 + 39? 49 + 8? 6 + 59? 69 + 3? 4 + 79? 89 + 2? 8 + 91?
15. How many are 27 + 8? 6 + 37? 47 + 4? 7 + 57? 67 + 8? 2 + 77? 87 + 9? 78 + 9? 8 + 79? 89 + 7?

LESSON LXVII.

1. I had 15 cents, and Charles gave me 5 more: how many cents had I then?

2. My slate cost 12 cents, and my primer, 10 cents: how much did both cost?

3. Mary paid 20 cents for a reader, and 5 cents for a pencil: how much did she pay for both?

4. Frank's coat cost 14 dollars, and his boots 5 dollars: how much did both cost?

5. Harry is 12 years old, and Susan is 9: how many years in both their ages?

6. Frank had 16 cents, and his aunt gave him 9 more: how many cents did he then have?

7. John owes me 13 cents, and Samuel, 10 cents: how much do both owe me?

8. I bought a whip for 18 cents: at what price must I sell it to make 6 cents?

9. Harvey is now 17 years old: in 10 years from this time, how old will he be?

10. In a school, there are 19 boys and 10 girls: how many pupils in the school?

11. Mary had 36 chickens, and she bought 4 more: how many had she then?

12. Oliver had 17 ducks, and his mother gave him 4 more: how many had he then?

13. Cora spent 47 cents for books, and 4 cents for pens: how much did she spend?

14. Edwin has 8 oranges more than Anna, and Anna has 19: how many has Edwin?

15. George bought a sled for 27 cents, and paid 5 cents to have it repaired: how much did the sled cost him?

16. Forty and 10 are how many?

LESSON LXVIII.

1. Thomas had 15 marbles, and lost 4: how many had he then?

2. Oscar had 16 cents, and spent 3: how many had he left?

3. Daniel, having 17 plums, gave his sister 4 of them: how many did he then have?

4. Charles bought 18 peaches, and gave 5 to a poor man: how many had he left?

5. Six and how many make 19? 7 and how many make 20?

6. If you have 20 cents, and spend 6, how many will you have left?

7. Sarah had 31 needles, and lost 2: how many had she left?

8. A boy had 33 chickens, and sold 4: how many had he remaining?

9. Lucy had 35 eggs, and broke 6 of them: how many had she then?

10. Henry had 10 cents, and his mother gave him enough more to make 40 cents: how much did she give him?

11. William had 56 cents, and spent all but 7 of them for school-books: how many cents did he spend?

12. Frank gathered 43 quarts of chestnuts: after selling part of them, he had only 8 quarts left: how many quarts did he sell?

13. Harry Lee owed me 53 cents: he has paid me 6 cents: how many cents does he yet owe me?

14. Thomas had 43 marbles, and gave 10 of them to his brother Charles: how many had he remaining?

15. How many are 65 less 9?

LESSON LXIX.

1. How many are $5 + 9 + 3 - 4$?

SOLUTION.—5 plus 9 plus 3 minus 4, equals 13.

2. How many are $6 + 10 + 4 - 3$?
3. How many are $4 + 9 + 5 - 6$?
4. How many are $7 + 4 + 10 - 3$?
5. How many are $10 + 9 + 8 - 4$?
6. How many are $9 + 8 + 8 - 6$?
7. How many are $4 + 9 + 7 - 6$?
8. How many are $9 + 6 + 8 - 7$?
9. How many are $9 + 7 + 8 - 4$?
10. How many are $18 + 8 + 10 - 9$?
11. How many are $9 + 9 + 9 - 3$?
12. How many are $13 + 10 + 3 - 4$?
13. How many are $17 + 9 + 2 - 7$?
14. How many are $11 + 10 + 9 - 5$?
15. How many are $21 + 8 + 8 - 7$?
16. How many are $19 + 9 + 6 - 5$?
17. How many are $10 + 8 + 6 - 9$?
18. How many are $18 + 9 + 10 - 6$?
19. How many are $21 + 9 + 10 - 8$?
20. How many are $17 + 10 + 8 - 9$?
21. How many are $20 + 10 + 4 - 6$?
22. How many are $30 + 6 + 4 - 7$?
23. How many are $39 + 4 + 8 - 6$?
24. How many are $47 + 6 + 9 - 7$?
25. How many are $54 + 9 + 10 - 8$?
26. How many are $50 + 3 + 7 - 4$?
27. How many are $68 + 8 + 10 - 6$?
28. How many are $80 + 4 + 5 - 6$?
29. How many are $84 + 4 + 8 - 7$?

LESSON LXX.

1. Frank had 19 cents, and spent 5: how many cents had he left?

2. Henry had 25 cents: he spent 4 cents for a top, and 6 cents for a kite: how many cents has he left?

3. Mother gave me 15 cents, and father gave me enough more to make 25 cents: how much did father give me?

4. I had 22 oranges: I gave 4 to brother Charles, and 6 to sister Mary: how many did I give away, and how many did I have left?

5. Thomas had 45 cents: he paid 5 cents for ink, and 10 cents for a copy-book: how many cents had he left?

6. My uncle gave me 35 cents: I bought a pen-knife for 20 cents, and a spelling-book for 10 cents: how many cents had I left?

7. Frank had 40 cents: he paid 10 cents for three oranges, and 18 cents for five lemons: how many cents has he remaining?

8. Charles is 4 years old, and his father is 32 years old: in how many years will Charles be as old as his father is now?

9. I paid 75 cents for a pair of skates, and 10 cents for a book: how much more did the skates cost than the book?

10. Harry had 40 cents given him on Christmas day: he spent 10 cents for toys, and 26 cents for books: how many cents had he remaining?

11. Mary had 50 cents: she gave 25 cents for a reader, 10 cents for a slate, and 5 cents for a sponge: how many cents did she pay for all, and how many had she left?

LESSON LXXI.

1. How many are $3 \times 4 - 5$?

SOLUTION.—3 multiplied by 4 equals 12, and 12 minus 5 equals 7

2. How many are $6 \times 6 - 7$? $7 \times 7 - 8$?
3. How many are $9 \times 9 - 10$? $3 \times 5 - 4$?
4. How many are $4 \times 5 - 6$? $3 \times 4 - 8$?
5. How many are $8 \times 5 - 10$? $6 \times 3 - 2$?
6. How many are $6 \times 8 - 4$? $6 \times 10 - 3$?
7. How many are $5 \times 10 - 7$? $9 \times 4 - 5$?
8. How many are $7 \times 6 - 3$? $4 \times 8 - 5$?
9. How many are $2 \times 10 - 5$? $9 \times 7 - 10$?
10. How many are $10 \times 3 - 8$? $8 \times 9 - 5$?
11. How many are $2 \times 3 - 4$? $4 \times 6 - 3$?
12. How many are $6 \times 2 - 5$? $3 \times 8 - 7$?
13. How many are $2 \times 4 - 8$? $9 \times 6 - 10$?
14. How many are $10 \times 7 - 2$? $2 \times 5 - 3$?

15. How many are $3 \times 9 - 7 + 1$?
16. How many are $10 \times 9 - 8 + 5$?
17. How many are $8 \times 8 - 9 + 2$?
18. How many are $3 \times 7 - 3 + 5$?
19. How many are $5 \times 6 - 9 + 4$?
20. How many are $5 \times 7 - 6 + 7$?
21. How many are $10 \times 4 - 6 + 3$?
22. How many are $7 \times 4 - 3 + 8$?
23. How many are $2 \times 9 - 7 + 9$?
24. How many are $5 \times 9 - 6 + 1$?
25. How many are $8 \times 7 - 9 + 6$?
26. How many are $2 \times 2 - 4 + 10$?
27. How many are $2 \times 7 - 9 + 5$?
28. How many are $10 \times 10 - 6 + 3$?

LESSON LXXII.

1. Four are how many times 2?

SOLUTION.—4 are 2 times 2.

2. Six are how many times 2? How many times 3?
3. Eight are how many times 2? How many times 4?
4. Nine are how many times 3?
5. Ten are how many times 2? How many times 5?
6. Twelve are how many times 2? How many times 3? 4? 6?
7. Fourteen are how many times 2? How many times 7?
8. Fifteen are how many times 3? How many times 5?
9. Sixteen are how many times 2? How many times 4? 8?
10. Eighteen are how many times 2? How many times 3? 6? 9?
11. Twenty are how many times 2? How many times 4? 5? 10?
12. Twenty-one are how many times 3? How many times 7?
13. Twenty-four are how many times 3? How many times 4? 6? 8?
14. Twenty-five are how many times 5?
15. Twenty-seven are how many times 3? How many times 9?
16. Twenty-eight are how many times 4? How many times 7?
17. Thirty are how many times 3? How many times 5? 6? 10?
18. Thirty-two are how many times 4? How many times 8?

LESSON LXXIII.

1. Thirty-five are how many times 5? How many times 7?
2. Thirty-six are how many times 4? How many times 6? 9?
3. Forty are how many times 4? How many times 5? 8? 10?
4. Forty-two are how many times 6? How many times 7?
5. Forty-five are how many times 5? How many times 9?
6. Forty-eight are how many times 6? How many times 8?
7. Forty-nine are how many times 7?
8. Fifty are how many times 5? How many times 10?
9. Fifty-four are how many times 6? How many times 9?
10. Fifty-six are how many times 7? How many times 8?
11. Sixty are how many times 6? How many times 10?
12. Sixty-three are how many times 7? How many times 9?
13. Sixty-four are how many times 8?
14. Seventy are how many times 7? How many times 10?
15. Seventy-two are how many times 8? How many times 9?
16. Eighty are how many times 8? How many times 10?
17. Eighty-one are how many times 9?
18. Ninety are how many times 9? How many times 10?
19. One hundred are how many times 10?

LESSON LXXIV.

PROMISCUOUS QUESTIONS.

1. George bought 2 peaches, at 3 cents each, and 2 oranges, at 5 cents each: how many cents did he pay for all?

2. Edwin has 4 oranges, and Thomas has 3 times as many as Edwin: how many oranges have both?

3. Anna is 6 years old, and Jane is twice as old as Anna, and 2 years more: how old is Jane? How many years in both their ages?

4. How many pine-apples, at 9 cents each, can you buy for 27 cents? for 45 cents? for 63 cents? for 72 cents?

5. I bought 10 cents worth of lemons, giving 2 cents for each lemon: how many lemons did I purchase?

6. If 3 men can do a certain piece of work in 5 days, how many men can do the same work in 1 day?

7. If a man can travel 90 miles in 9 hours, how many miles can he travel in 1 hour? In what time could he travel 20 miles?

8. Samuel bought 3 books, at 10 cents each, and a toy for 7 cents: how many cents did he pay for all?

9. If 4 men can do a certain piece of work in 9 days, in how many days can 1 man do the same work?

10. If 10 peaches are worth 1 orange, how many oranges are 60 peaches worth? How many are 80 peaches worth? 100 peaches?

11. I bought 3 pounds of raisins, at 8 cents a pound, and 2 oranges for 10 cents: how much did all cost?

12. What will 8 cords of wood cost, at seven dollars a cord?

LESSON LXXV.

1. Charles and Henry had each 10 marbles: Charles gave 6 of his to Henry: how many did each have then?

2. William Jones owed me 20 cents: he gave me 3 peaches, worth 4 cents each, and an orange, worth 5 cents: how much was then due?

3. I bought 3 oranges, at 5 cents each, and 2 lemons, at 4 cents each: how many cents did I pay for all?

4. When 3 lemons were selling for 15 cents, John gave 1 lemon and 5 cents in money for a book: what was the value of the book?

5. Harry bought 3 rabbits for 30 cents, and sold them for 39 cents: how many cents did he gain on the 3 rabbits? On each rabbit?

6. I bought 3 dozen eggs, at 6 cents a dozen, and sold them at 8 cents a dozen: how much did I make?

7. Frank bought 4 flags, at 10 cents each, and 3, at 3 cents each: how many flags did he buy, and how much did they cost?

8. John gave 30 cents in money and 3 peaches, worth 5 cents each, for a sled: how much did the sled cost?

9. I paid 25 cents for five pounds of meat, and 10 cents for a melon: how much did I pay for all?

10. Mary bought 3 quarts of chestnuts, at 10 cents a quart, and a doll for 20 cents: how much did all cost?

11. Frank bought 5 books, at 7 cents each, and sold them at 9 cents each: how many cents did he make?

12. I bought 3 dozen pens, at 10 cents a dozen, and some paper for 18 cents: how much did all cost me?

13. A drover bought 10 sheep, at 3 dollars a head, and sold them for 5 dollars a head: how much did he gain?

LESSON LXXVI.

1. James bought 3 lemons, at 2 cents each, and paid for them with oranges, at 3 cents each: how many oranges did it take?

2. Mary bought 3 yards of ribbon, at 4 cents a yard: how many cents did it cost? She paid for it with two-cent coins: how many coins did it take?

3. Daniel bought 8 tops, at 2 cents each, and paid for them with oranges, at 4 cents each: how many oranges were required?

4. Francis bought 9 marbles, at 2 cents each, and paid for them with tops, at 3 cents each: how many tops did it take?

5. Sarah bought 4 thimbles, at 5 cents each, and paid for them with cherries, at 10 cents a quart: how many quarts did she give?

6. A man bought 8 yards of cloth, at 3 dollars a yard, and paid for it with flour, at 4 dollars a barrel: how many barrels did it take?

7. Thomas bought 3 oranges, at 5 cents each, and paid for them with chestnuts, at 3 cents a quart: how many quarts did he give?

8. I bought 4 barrels of flour, at 10 dollars a barrel, and paid for it with apples, at 5 dollars a barrel: how many barrels of apples did I give?

9. I bought 3 pounds of raisins, at 8 cents a pound, and paid for them with melons, at 6 cents each: how many melons did I give?

10. How many oranges, at 6 cents each, will pay for 10 lemons, at 3 cents each?

11. How many pears, at 8 cents each, would it take to buy 4 pine-apples, at 10 cents each?

LESSON LXXVII.

1. How many pine-apples, at 10 cents each, will pay for 5 peaches, at 6 cents each?

2. How many pencils, at 5 cents each, will pay for 4 books, at 10 cents each?

3. How many kites, at 4 cents each, will pay for 6 lemons, at 6 cents each?

4. How many peaches, at 5 cents each, will pay for 4 slates, at 10 cents each?

5. I bought 3 quarts of strawberries, at 10 cents a quart, and paid for them with chestnuts, at 6 cents a quart: how many quarts did it take?

6. I bought 5 dozen figs, at 8 cents a dozen, and paid for them with pears, at 4 cents each: how many pears did I give?

7. If 4 men can mow a field of grass in 5 days, in how many days can 10 men mow the same field?

8. How much cloth, at 6 dollars a yard, will pay for 4 barrels of flour, at 9 dollars a barrel?

9. How many slates, at 3 dimes each, will cost as much as 2 geographies, at 6 dimes each?

10. How many bottles of ink, at 8 cents each, will pay for 10 oranges, at 4 cents each?

11. How many oranges, at 4 cents each, will pay for 2 books, at 10 cents each?

12. If 6 men can do a piece of work in 8 days, how long will it take 4 men to perform it?

13. If 7 barrels of flour cost 84 dollars, what will 5 barrels cost?

14. An orchard contains 8 rows of trees, and has 6 trees in each row; if the same number of trees were placed in 4 rows, how many would there be in a row?

LESSON LXXVIII.

1. Begin with 2, and add by 2's to 100.
2. Begin with 100, and subtract by 2's to 0.
3. Begin with 3, and add by 3's to 99.
4. Begin with 99, and subtract by 3's to 0.
5. Begin with 4, and add by 4's to 100.
6. Begin with 100, and subtract by 4's to 0.
7. Begin with 5, and add by 5's to 100.
8. Begin with 100, and subtract by 5's to 0.
9. Begin with 6, and add by 6's to 96.
10. Begin with 96, and subtract by 6's to 0.
11. Begin with 7, and add by 7's to 98.
12. Begin with 98, and subtract by 7's to 0.
13. Begin with 8, and add by 8's to 96.
14. Begin with 96, and subtract by 8's to 0.
15. Begin with 9, and add by 9's to 99.
16. Begin with 99, and subtract by 9's to 0.
17. Begin with 1, and add by 2's to 99.
18. Begin with 100, and subtract by 3's to 1.
19. Begin with 1, and add by 3's to 100.
20. Begin with 97, and subtract by 4's to 1.
21. Begin with 1, and add by 4's to 97.
22. Begin with 96, and subtract by 5's to 1.
23. Begin with 1, and add by 5's to 96.
24. Begin with 97, and subtract by 6's to 1.
25. Begin with 1, and add by 6's to 97.
26. Begin with 100, and subtract by 6's to 4.
27. Begin with 1, and add by 7's to 99.
28. Begin with 100, and subtract by 7's to 2.
29. Begin with 1, and add by 8's to 97.
30. Begin with 100, and subtract by 8's to 4.
31. Begin with 1, and add by 9's to 100.

TABLES.

LESSON LXXIX.

UNITED STATES MONEY.

10 cents (cts.)	make	1 dime :	marked d.
10 dimes	"	1 dollar :	" $.
10 dollars	"	1 eagle :	" E.

1. How many cents in 3 dimes?

SOLUTION.—In 1 dime there are 10 cents; then, in 3 dimes there are 3 times 10 cents, which are 30 cents.

2. How many dimes in 3 dollars? In 7 dollars?
3. How many dimes in 60 cents?

SOLUTION.—In 10 cents there is 1 dime; then, in 60 cents there are as many dimes as 10 is contained times in 60, which are 6.

4. How many dollars in 6 eagles? In 8? In 4?
5. How many dollars in 40 dimes? In 60? In 90?
6. How many dimes in 1 eagle? In 8 dollars?
7. If 1 lemon cost 6 cents, how many dimes will 5 lemons cost?
8. James bought 8 pencils, at 5 cents each: how many dimes did they cost?
9. How many dollars will pay for 5 bushels of wheat, at 10 dimes a bushel?
10. Jane spent 10 cents in one store, 3 dimes in another, and 5 dimes in another: how many cents did she spend altogether? How many dimes?
11. Henry spent 6 dimes for handkerchiefs, 25 cents for collars, and gave 1 dollar to the clerk in payment: how many cents in change did he receive?

LESSON LXXX.

ENGLISH MONEY.

4 farthings (far.)	make	1 penny	: marked d.
12 pence	"	1 shilling	: " s.
20 shillings	"	1 pound	: " £.

1. How many farthings in 3 pence?

SOLUTION.—In 1 penny there are 4 farthings; then, in 3 pence there are 3 times 4 farthings, which are 12 farthings.

2. How many shillings in 60 pence?

SOLUTION.—In 12 pence there is 1 shilling; then, in 60 pence there are as many shillings as 12 is contained times in 60, which are 5.

3. How many pence in 6 shillings? In 8? In 7?

4. How many shillings in 2 pounds? In 4? In 5?

5. How many pounds in 40 shillings? In 60? In 100?

6. If a geography cost 4 shillings, how many pounds will 10 geographies cost?

7. A boy had 9 pence; his father gave him enough to make 2 shillings: how much did he receive.

8. How many books, at 6 shillings each, can be bought for 3 pounds?

9. How many tops, at 3 pence each, can be bought for 2 shillings?

10. If 10 yards of cloth costs one pound, how many shillings does 1 yard cost?

11. A boy in London spent 12 pence in one store, 12 shillings in another, and 7 shillings in another: how many pounds did he spend?

12. Jane spent 8 pence for prunes, 16 pence for rice, and 5 shillings for sugar, giving the clerk 1 pound in payment: how many shillings in change did she receive?

LESSON LXXXI.

TROY WEIGHT.

24 grains (gr.)	make	1 penny-weight	: marked	pwt.
20 penny-weights	"	1 ounce	: "	oz.
12 ounces	"	1 pound	: "	lb.

1. How many ounces in 5 pounds?

SOLUTION.—In 1 pound there are 12 ounces; then, in 5 pounds there are 5 times 12 ounces, which are 60 ounces.

2. How many pounds in 96 ounces?

SOLUTION.—In 12 ounces there is 1 pound; then, in 96 ounces there are as many pounds as 12 is contained times in 96, which are 8.

3. How many penny-weights in 3 ounces? In 5?
4. How many ounces in 60 penny-weights? In 80?
5. How many grains in 2 penny-weights?
6. How many penny-weights in 48 grains?
7. If 1 penny-weight of silver is worth 3 dimes, how many dimes is 1 ounce worth?
8. A miner had 40 penny-weights of gold in one bag, 6 ounces in another, and 4 ounces in another: how much did he have in all?
9. If an ounce of gold is worth £3, what is 1 pound of gold worth?
10. Bought a gold chain for $20, paying 2 dollars a penny-weight for it: how much did it weigh?
11. If 1 penny-weight of gold is worth 3 penny-weights of silver, how many ounces of silver are worth 1 pound of gold?
12. A man owed 3 pounds of silver: at one time he paid 14 ounces; at another, 20 ounces; and at another, 20 penny-weights: how many ounces did he still owe?

LESSON LXXXII.

AVOIRDUPOIS WEIGHT.

16 ounces (oz.)	make	1 pound	:	marked	lb.
100 pounds	"	1 hundred-weight	:	"	cwt.
2000 pounds	"	1 ton	:	"	T.

1. How many ounces in 2 pounds? In 3?
2. How many pounds in 32 ounces? In 64?
3. How many ounces in 4 pounds? In 5?
4. How many pounds in 48 ounces? In 80?
5. How many pounds in 2 tons? In 3?
6. How many tons in 4000 pounds? In 6000?
7. I bought 20 ounces of rice in one store, and 12 ounces in another: how many pounds did I buy at both stores?
8. James bought 3 pounds of brown sugar and 1 pound of white sugar: how many ounces did he buy?
9. A farmer sold 4000 pounds of hay to one man, and 8000 pounds to another: how many tons did he sell?
10. Frederick caught one fish that weighed 48 ounces, and another that weighed 32 ounces: how many pounds did the two fishes weigh?
11. I bought 1 pound of figs for 20 cents, and sold them for 3 cents an ounce: how much did I get for my figs? How many cents did I make?
12. An express company received 4 barrels of fish, weighing 1 hundred-weight each, and charged 2 cents a pound for transportation: how much were their charges?
13. How many pounds in 6 cwt.?
14. How many pounds in 5 tons?
15. If you buy 5 pounds of coffee, how many ounces do you get?

LESSON LXXXIII.

DRY MEASURE.

2 pints (pt.)	make	1 quart	:	marked	qt.
8 quarts	"	1 peck	:	"	pk.
4 pecks	"	1 bushel	:	"	bu.

1. How many pints in 2 quarts? In 3? In 4?
2. How many quarts in 12 pints? In 14? In 16?
3. How many pecks in 2 bushels? In 3? In 5?
4. How many bushels in 16 pecks? In 24? In 32?
5. How many pints in 2 pecks? In 3?
6. How many quarts in 2 bushels? In 3?
7. If 1 pint of meal cost 2 cents, what will 7 quarts cost?
8. If 3 pecks of apples cost 24 cents, what will 1 quart cost?
9. I bought 10 quarts of oats of one man, 9 of another, and 5 of another: how many pecks did I buy?
10. Henry bought 2 bushels of peaches, and sold 6 pecks of them: how many quarts of peaches had he left?
11. James bought 3 bushels of apples for 6 dollars, and sold them for 1 dollar a peck: how much did he get for his apples: how many dollars did he make?
12. Henry bought 6 quarts of berries for 30 cents, and 10 quarts for 40 cents: how many pecks did he buy? How much did all the berries cost?
13. A farmer bought 12 bushels of seed for 60 dollars: he sold 35 dollars worth to a neighbor at the same price per bushel that he gave: how many bushels did he sell?
14. If I feed my horse 4 quarts of oats three times a day: how many pecks will be required for six days?

LESSON LXXXIV.

LIQUID OR WINE MEASURE.

4 gills (gi.)	make	1 pint	: marked	pt.
2 pints	"	1 quart	: "	qt.
4 quarts	"	1 gallon	: "	gal.

1. How many gills in 4 pints? In 6? In 8?
2. How many pints in 20 gills? In 28? In 36?
3. How many pints in 6 quarts? In 8? In 10?
4. How many quarts in 12 pints? In 16? In 18?
5. How many quarts in 5 gallons? In 7? In 9?
6. How many gallons in 16 quarts? In 24? In 32?
7. If 6 gills of olive-oil cost 60 cents, what will 1 pint cost?
8. If 8 quarts of milk cost 48 cents, what will 1 pint cost?
9. At 80 cents a gallon, what will 1 pint of molasses cost?
10. My cook buys 8 quarts of milk every week, at 4 cents a pint. How much does the milk cost a week?
11. I bought 6 quarts of vinegar at one time, 7 at another, and 11 at another: how many gallons did I buy?
12. Frederick bought 4 gallons of milk, at 2 dollars a gallon: he sold it at 1 dollar a quart: how much did he make on his milk?
13. A grocer bought 10 gallons of kerosene for 3 dollars, and sold it for 40 cents a gallon: how much did ho make on each gallon?
14. How many quarts in 30 pints of milk?
15. If a gallon of syrup costs 1 dollar, how many pints do you get for 5 dollars?

LESSON LXXXV.

LONG MEASURE.

12	inches (in.)	make 1 foot	: marked	ft.
3	feet	" 1 yard	: "	yd.
$5\frac{1}{2}$	yards (or $16\frac{1}{2}$ feet)	" 1 rod	: "	rd.
320	rods	" 1 mile	: "	mi.

1. How many inches in 3 feet? In 5? In 7?
2. How many feet in 24 inches? In 48? In 96?
3. How many feet in 4 yards? In 8? In 10?
4. How many yards in 9 feet? In 15? In 27?
5. How many rods in 2 miles? In 3?
6. One desk is 2 feet in length, another 3 feet: how many inches longer is one desk than the other?
7. A man made a sidewalk 40 feet long, and a walk to the front door 14 feet long: how many yards of walk did he make?
8. Robert put weather-strips on 5 windows, 4 yards being required for each window: how many feet of weather-strips did he put on all the windows?
9. I have one piece of cloth, 1 yd. and 10 in. in length, another is 26 inches in length: how many yards of cloth in both pieces?
10. John can jump 50 inches, and his brother can jump ten inches farther than John: how many feet can his brother jump?
11. How many bricks, each 8 inches long, will it take to make a single course of bricks 2 yards long?
12. A stairway consists of 12 steps, each 8 inches in height: what is the entire rise in feet?
13. How many miles in 960 rods?
14. How many inches in 10 yards?

LESSON LXXXVI.

SQUARE MEASURE.

144	square	inches	make	1 square foot	:	marked	sq. ft.
9	"	feet	"	1 " yard	:	"	sq. yd.
30¼	"	yards	"	1 " rod	:	"	sq. rd.
160	"	rods	"	1 acre	:	"	A.
640	acres		"	1 square mile	:	"	sq. mi.

1. How many sq. ft. in 3 sq. yds.? In 5?
2. How many sq. yds. in 36 sq. ft.? In 72?
3. How many sq. rds. in 2 A.? In 3?
4. How many A. in 320 sq. rds.? In 480?
5. How many sq. ft. in a blackboard 9 feet long and 3 feet wide?
6. A floor is 5 yds. long and 4 yds. wide: how many sq. yds. does it contain?
7. One closet measures 36 sq. ft., and another 45 sq. ft.: how many sq. yds. in both closets?
8. Jane wishes to carpet a room that measures 15 feet in length and 12 feet in width: how many yards of carpet, 1 yard wide, must she buy?
9. How much will it cost to plaster a wall 6 feet high and 9 feet long, at 7 cents a sq. yd.?
10. A fern-case, 2 feet square, rests on a stand 1 yard square: how many sq. ft. of the surface of the stand are not occupied by the fern-case?
11. How many square feet in a walk that is 1 yard wide and 4 yards long?
12. How many square feet of lumber in a board that is 24 inches wide and 16 feet long?
13. How many square miles in 1280 acres?
14. How many square feet in 10 square yards?

LESSON LXXXVII.

APOTHECARIES WEIGHT.

20 grains (gr.)	make	1 scruple	: marked	℈.
3 scruples	"	1 dram	: "	ʒ.
8 drams	"	1 ounce	: "	℥.
12 ounces	"	1 pound	: "	℔.

SOLID OR CUBIC MEASURE.

1728 cubic inches	make	1 cubic foot	: marked	cu. ft.
27 cubic feet	"	1 cubic yard	: "	cu. yd.
128 cubic feet	"	1 cord.		

NOTE.—A pile of wood, 8 feet long, 4 feet wide, and 4 feet high contains a cord.

TIME TABLE.

60 seconds (sec.)	make	1 minute	: marked	min.
60 minutes	"	1 hour	: "	hr.
24 hours	"	1 day	: "	da.
365 days, 6 hours ($365\frac{1}{4}$ da.)	"	1 year	: "	yr.
100 years	"	1 century	: "	cen.

7 days	make	1 week	: marked	wk.
4 weeks	"	1 month	: "	mo.
12 calendar months	"	1 year	: "	yr.

The following table gives the number of days in each month:

1. January,	31	days.
2. February,	28 or 29	"
3. March,	31	"
4. April,	30	"
5. May,	31	"
6. June,	30	"
7. July,	31	days.
8. August,	31	"
9. September,	30	"
10. October,	31	"
11. November,	30	"
12. December,	31	"

LESSON LXXXVIII.

CIRCULAR MEASURE.

60 seconds (″)	make	1 minute :	marked ′.
60 minutes	"	1 degree :	" °.
360 degrees	"	1 circle.	

PAPER.

24 sheets of paper	make	1 quire.
20 quires " "	"	1 ream.

BOOKS.

When the sheets of paper in a book are so folded that each sheet makes 2 leaves, or 4 pages, the book is called a folio, fol.

4 leaves, or	8 pages,	the book is called	a quarto,	4to.
8 "	" 16	" " " " "	an octavo,	8vo.
12 "	" 24	" " " " "	a duodecimo,	12mo.
16 "	" 32	" " " " "	a	16mo.
18 "	" 36	" " " " "	an	18mo.

MISCELLANEOUS TABLES.

12 things	make	1 dozen.	6 feet	make	1 fathom.
12 dozen	"	1 gross.	4 inches	"	1 hand.
20 things	"	1 score.	4 rods	"	1 chain.

---

196 pounds of flour	make	1 barrel.
200 pounds of beef or pork	"	1 barrel.
60 pounds of wheat	"	1 bushel.
56 pounds of corn	"	1 bushel.

LESSON LXXXIX.

REVIEW OF TABLES.

1 How many cents in 5 dimes? In 3? In 7?
2. How many pence in 40 farthings? In 32?
3. How many eagles in 30 dollars? In 50? In 70?
4. How many shillings in 3 pounds? In 5? In 4?
5. How many penny-weights in 4 ounces? In 2?
6. How many pounds Troy in 36 ounces? In 60?
7. How many ounces avoirdupois in 2 pounds? In 3?
8. How many pounds avoirdupois in 32 ounces? In 48?
9. How many quarts in 6 pecks? In 8? In 10?
10. How many bushels in 20 pecks? In 28? In 36?
11. How many quarts in 7 gallons? In 6? In 8?
12. How many gallons in 16 quarts? In 20? In 36?
13. How many feet in 7 yards? In 10? In 12?
14. How many square yards in 45 square feet? In 63?
15. How many scruples in 9 drams? In 8? In 10?
16. How many cubic feet in 2 cubic yards?
17. How many cubic yards in 54 cubic feet?
18. How many hours in 2 days? In 3?
19. How many days in 48 hours? In 72?
20. How many weeks in 7 months? In 12?
21. How many minutes in 2 degrees? In 3?
22. How many quires in 4 reams of paper? In 3?
23. How many pens in 1 gross?
24. How many feet in 7 fathoms? In 5? In 9?
25. How many feet tall is a horse 15 hands high?
26. How many chains in 24 rods? In 28? In 36?
27. How many farthings in 2 pounds? In 10?
28. How many ounces in 80 penny-weights?
29. How many sheets of paper in 2 quires? In 2 reams?
30. How many are 3 score and 10?

Who Is Joseph Ray?

Joseph Ray lived as a contemporary of Abraham Lincoln. Youth in that generation finished their schoolbooks and then read the Bible, sang from the hymnbooks of Lowell Mason, and read Roman and Greek classics in the original languages. It was not unusual for a blacksmith to carry a Greek New Testament under his cap for reading during his lunch break. The literacy rate, even on the frontier, was higher than today's rate.

Ray was Professor of Mathematics for twenty-five years at a preparatory school in Ohio. He had no use for indolence and sham. He was always delighted to join his students in sports. He knew how to use balls, marbles, and tops as concrete illustrations to help young children make the transfer from solid objects to abstract figures.

From the Presidency of Abraham Lincoln to that of Teddy Roosevelt, few Americans went to school or were taught at home without considerable exposure to either Ray's Arithmetics or McGuffey's Readers—usually both. Ray and McGuffey challenged students to excellent accomplishment. Their influence on our country has certainly eclipsed Mann's and rivaled Dewey's, yet education histories, edited by humanists, seldom mention these men.

Ray's classic Arithmetics are now brought to a new generation which is in search of excellence.

Ray's Arithmetic Series

Primary Arithmetic. Reading, writing and understanding numbers to 100; adding and subtracting with sums to 20; multiplication and division to 10s; and signs and vocabulary needed for this level of arithmetic.

Intellectual Arithmetic. Reading, writing and understanding of higher whole numbers, fractions, and mixed numbers; addition, subtraction, multiplication and division of higher numbers; computation of simple fractions; beginning ratio and percentage; and signs and vocabulary needed for all these operations.

Practical Arithmetic. Roman numbers; carrying in addition and borrowing in subtraction; measurement and compound numbers; factors; decimals and percentage; ratio and proportion; powers and roots; beginning geometry; advanced vocabulary.

Higher Arithmetic. Philosophical understandings; principles and properties of numbers; advanced study of common and decimal fractions, measurements, ratio, proportion, percentage, powers, and roots; series; business math; geometry.

Test Examples. A supply of problems for making tests to accompany study in *Practical Arithmetic* and *Higher Arithmetic*.

Key to Ray's Primary, Intellectual and Practical Arithmetics. Answers to problems in the three lower books.

Key to Ray's Higher Arithmetic. Answers to problems in the higher book.

Parent-Teacher Guide. Gives unit by unit helps for teaching; suggests grade levels for each book; provides progress chart samples for each grade and tests for each unit.